INSECT SAFARI

Exploring the Wondrous World of Everyday Bugs

MARGIE PATLAK

WORKMAN PUBLISHING | NEW YORK

Copyright © 2026 by Margie Patlak
Photos copyright © 2026 by Margie Patlak, except for those on page 268

Hachette Book Group supports the right to free expression and the value of copyright. The purpose of copyright is to encourage writers and artists to produce the creative works that enrich our culture.

The scanning, uploading, and distribution of this book without permission is a theft of the author's intellectual property. If you would like permission to use material from the book (other than for review purposes), please contact permissions@hbgusa.com. Thank you for your support of the author's rights.

Workman
Workman Publishing
Hachette Book Group, Inc.
1290 Avenue of the Americas
New York, NY 10104
workman.com

Workman is an imprint of Workman Publishing, a division of Hachette Book Group, Inc. The Workman name and logo are registered trademarks of Hachette Book Group, Inc.

Design by Katie Benezra
Cover art by Lucy Rose
Cover design by Reagan Ruff and Katie Benezra

Similar versions of "Marveling at Dragonfly Superheroes" and "Unearthing Superorganisms" were previously published in the online magazine *Zygote Quarterly*.

The publisher is not responsible for websites (or their content) that are not owned by the publisher.

Workman books may be purchased in bulk for business, educational, or promotional use. For information, please contact your local bookseller or the Hachette Book Group Special Markets Department at special.markets@hbgusa.com.

Library of Congress Cataloging-in-Publication Data is on file.

ISBN 978-1-5235-3306-0 (paperback), 978-1-5235-3307-7 (ebook)

First Edition April 2026

Printed in China (1010) on responsibly sourced paper

Cover © 2026 Hachette Book Group, Inc.

10 9 8 7 6 5 4 3 2 1

Instructions for living a life:
Pay attention.
Be astonished.
Tell about it.

—MARY OLIVER

It is easy to forget that human beings form a tiny two-legged minority in an overwhelmingly six-legged world.

—STEPHEN MARSHALL

When you don't understand the abilities of a species, you lose something that's part of the world, and part of everything around you.

—IRENE PEPPERBERG

Contents

• A fritillary drinks nectar from a purple coneflower.

Introduction

It was the cutest insect I'd ever seen.

As the tricolored bumble bee dipped into the backlit tunnel of a petunialike calibrachoa flower to sip some nectar, I glimpsed bright orange pantaloons of pollen on her rear legs. A rust-colored belt was buried between pale yellow and black furry stripes on her stout body. The bee's fur gave her an endearing teddy-bear look, something you could envision cuddling with—not the case for most insects.

Maybe I gave the bee that laudatory label of "cutest" because it was one of the first insects I had ever looked at closely and then tried to identify. I had a degree in environmental studies and wrote a book about the nature in Maine, covering everything from its rocks and clouds to its moose and foxes, but up until that point, most insects had flown under my radar. Perhaps because they're so small and inconspicuous? Or because the two insects I knew best—mosquitoes and yellow jackets—did not endear this large category of animals to me?

I had always taken insects for granted, their hum the background hymn of summer. But then there was talk of a lurking insect apocalypse, and I realized how little I knew about this vast but rapidly shrinking number of beings omnipresent in nature. That's when it registered: If I were more mindful of insects, not only would my world expand immensely, but I would be more inspired to protect them.

• A tricolored bumble bee entering a *Calibrachoa* flower.

You can't protect what you don't know exists.

I started taking regular insect safaris, both in my tiny courtyard in the heart of Philadelphia and on the more extensive property I have in Down East Maine, where I live during most of the growing season. My home there is lodged in a twenty-acre evergreen forest, and its backyard is filled with flower beds bordering a stone pathway that winds its way to a rocky, muddy, marshy tidal beach. This diverse ecosystem attracts lots of different kinds of insects. But even during the rest of the year, at my Philadelphia townhouse, I see numerous insects drawn to the flowers I set out in pots.

Armed with only my cell phone camera and an observant and patient eye, I began to trail whatever flying or crawling bug came my way and snapped a photo when it stopped to rest or eat. I zoomed into the photos to get a closer look at those specks skating the surface of flower petals. I then found their names using the app iNaturalist, which deploys artificial intelligence combined with crowdsourcing by amateur and expert entomologists to identify species.

In just a few months' time, I had discovered many bizarre-looking insects: bird-poop-mimicking moths, flower flies copying bees' stripes, iridescent beetles and wasps, a large moth resembling a hummingbird, and millimeter-sized humpbacked dance flies wearing golden slippers of pollen. One year out, I had tallied more than two hundred different insect species, which isn't surprising considering that a typical backyard may contain several thousand species of insects and several million individual bugs.

Many are stunningly beautiful.

One morning I spotted something glinting on a leaf and discovered it was a 2-millimeter golden long-legged fly. Another time cropping one of my photos transformed a green speck that landed on my sweatshirt into an elongated glittering emerald tiger beetle with distinctive white spots and

• After using my phone camera to magnify something glinting on a leaf in my Philadelphia courtyard, I discovered it was a golden long-legged fly.

long antennae. Often the insects were more gorgeous than the flowers on which they landed.

The variety of butterflies, bees, and wasps I detected multiplied once I learned to carefully follow them until they stopped to dine on a flower. While they were immobilized there for a few seconds, I captured them with my camera and identified them later. I learned that orange-and-black

butterflies are not all monarchs, but rather several different species with fanciful names like great spangled fritillary, northern crescent, and painted lady. The bees in my garden weren't just the large and obvious bumble bees, of which I saw several kinds, but also the much smaller sweat, leafcutting, and longhorned bees. These smaller bees are only between one-quarter and one-half inch in length and barely noticeable with the naked eye. And there were so many different kinds of wasps, including some that were a surprising metallic turquoise color or iridescent cobalt blue. The wasps' sizes ranged from several inches long to so tiny I never saw them but suspected they were there by the distinctive holes they left behind.

I even discovered insects I hadn't realized were insects—it turns out that the half-inch-long, oblong patches of flat black scales attached to one of the walls of my house were the juvenile forms of fireflies, and the fat white "worms" with brown heads I found while digging in the garden were the larval forms of June bugs.

One morning I noticed a flower fly unusually still on a black-eyed Susan petal. I zoomed in on it with my phone's camera, and only then did I discover that the fly was in the clutches of a well-camouflaged crab spider. The spider

• A flower fly clutched by a crab spider, which injects its venom into the fly to paralyze and consume it.

had paralyzed the flower fly with its venom in order to consume it. It was a life-and-death drama befitting a David Attenborough nature show—all happening on a petal! I realized that I didn't have to go on safari in some faraway country to see such action and new, exotic creatures. Instead, I could venture out into my own backyard or any other place where there's a bit of nature and find many animals I had never seen before—hiding in plain sight.

Once I made the acquaintance of all these new creatures, I became curious about what their lives were like. What did they eat, where did they go in winter, how did they find their mates and raise their young? I had *lots* of questions. I scoured the internet for answers, read a bunch of books and scientific papers, took a few classes, and had many correspondences and chats with entomologists.

What I discovered flabbergasted me.

Because let's face it—insects are weird. First, they all have skeletons on the outside of their bodies, which is strange in itself. But their anatomy has even more surprises. Some can have lanterns in their bellies, while others have drums. Their taste buds can be on their legs and feet, and their ears nearly anywhere, including on the torso, legs, mouth, or even wings. Their eyes need not be on their heads—Japanese yellow swallowtail butterflies sport primitive eyes on their genitals, of all places! (These eyes guide the male's sexual apparatus over a female's and help the female position her egg-laying tube on the surface of a plant.)

Many common insects are also superheroes, with special powers, sensors, and secret codes. The diabolical ironclad beetle has a body frame so strong it can withstand being run over by a car. A leafhopper as tiny as a pinhead travels thousands of miles south each winter. A lacewing, though less than half an inch long and with wings so delicate that it is named for them, can withstand temperatures way below zero. Many insects have senses we lack, including those for detecting humidity, carbon dioxide, heat, airflow, water flow, surface vibrations, and smoke. Some have bodies that can sense and emit ultrasonic pulses. Others have eyes that can see many more colors than we can, and many more frames per second. Nearly all emit chemicals that can communicate across far distances in a secret species-specific language.

Also intriguing was the plethora of recent scientific findings revealing there's much more to the inner lives and behaviors of insects than most of us could imagine. Insects are having their moment. Who knew that wasps use tools and recognize faces, bees play with balls and do math, ants invented farming way before we did, and even fruit flies mull over their mating choices? These surprising findings reinforce the notion that we aren't the only intelligent beings on Earth and reveal alien-to-us life right here on our own planet.

But perhaps even more astonishing are the changes many insects undergo during their lifetimes. Unlike most animals, whose immature forms resemble those of their adult parents, grubs look nothing like the beetles or flies they become. Caterpillars are a far cry from butterflies. And the nymphs that dart in the water are poor second cousins to the shimmering dragonflies helicoptering above them. Through an all-encompassing metamorphosis, an insect often enters an entirely new domain: aquatic to terrestrial; below- to aboveground; crawling to soaring. It was strange enough for me as an adolescent girl to suddenly find myself with breasts and hips, so I can't even imagine how startling it would be to sprout wings and fly in the sky after spending years worming underground. The animal kingdom labored for millions of years, evolving through zillions of genetic glitches, trials and errors, natural selections, and extinctions and modifications, all leading to the extraordinary morphological changes that created the creatures we see today. But insects? They do all this effortlessly, overnight, *all the time.*

In fact, insects are more evolved than us humans, with lineages stretching much farther back in time. Such long-term evolution crafted many of their mind-blowing features, like being able to see using only starlight and navigate by the position of the sun. That long-term evolution reinforces the notion that insects are incredibly resilient. Surviving major cataclysms in our planet's history, from volcanic eruptions to asteroid collisions, continental smashups, and climate catastrophes, they made their mark in almost every era.

"I didn't have to go on safari in some faraway country to see new, exotic creatures. Instead, I could venture out into my own backyard or any other place where there's a bit of nature and find many animals I had never seen before—hiding in plain sight."

Due to their fantastic ability to adapt, the late biologist E. O. Wilson claimed that insects would inherit the world if humankind disappeared; but what is *their* world like now? After all, insects comprise four out of every five species on Earth. They can be found in droves in almost every nook and cranny of the planet, from frigid to salty to searing, such that "the Earth belongs more to the insects than to us," asserts entomologist Michael Engel in his book *Innumerable Insects*. Yet despite their abundance and omnipresence, most of us are acquainted with only a handful of insects, and know very little about even these few. We know where bears live in winter, where birds nest, and what lions eat, yet few of us can answer such questions about the insects all around us.

Exploring the amazing superpowers, alien anatomies, and unusual lifestyles of insects exposed the narrow limits of my own human existence and tuned me in to other possibilities. Like a Renaissance explorer who just found the strange new world of the Americas, I felt compelled to shout out all I was discovering so more people would notice and appreciate the bugs in their own backyards, flowerpots, and nearby parks instead of swatting them away—so that these remarkable little creatures, ever-present yet largely unnoticed in the world around us, would fill them with awe.

Insect Safari is my shout-out. It is chock-full of stories about some of the incredible insect characters I've personally come to know and appreciate, and the philosophical and moral musings they've inspired: whether maternal instinct is truly instinctual, the value of a short life, whether it's better to blend in or alter your environment, and whether it's possible to be altruistic in caring for your own kin without committing atrocities against those you might consider Other.

At first, I was intrigued and a little put off by how strange insects are, but the more I learned about them, the more I wanted to understand them. As author Sabrina Imbler pointed out in a *New York Times* essay, "Instead of gawking at such creatures as bizarre, I find it more fulfilling to seek connection with them across and because of our differences. . . . Nature is beautiful *and* full of weirdos. . . . Perhaps dwelling on these differences can incite wonder—a reminder of how many strange lives, bodies and ways of being there are on this planet." Echoing this sentiment in an essay for *Orion* magazine, journalist and podcaster Lulu Miller noted, "You may observe [insects] as the foreign, separate beings that they are, or you can root around in their exoskeletons for meaning, for metaphor, for some silky strand of resonance, for some kernel with which to recalibrate your compass."

Get ready to recalibrate your compass and go on safari with me. I promise that what you learn in these pages will expand your world.

I've organized *Insect Safari* into chapters based on some of the specific bugs I've encountered on my explorations in nature. I've also included a few

• This longhorn bee often escapes notice because it is so tiny—less than half an inch in length.

chapters, such as "Reading Insect Lore in Leaves," "Reflecting on Insects in Winter," and "Unmasking Insects Blending In," that encompass a general feature about insects or one shared by a large group of them. Although many of these chapters stand alone, most build on the knowledge revealed in previous chapters, so it's best to read them in order.

Let's go. There's a lot to discover out there.

TRAILING THE VERY BUSY BUMBLE BEE

Tricolored Bumble Bees (*Bombus ternarius*)

LOCATION: On a tansy (*Tanacetum vulgare*) flower in a garden, northeast Maine

DATE: August 23 TIME: 10:22 a.m.

The transformation was incredible—suddenly, everywhere I looked I saw bees!

Ironically, this happened after I learned about a drastic decline in the bee population. That's when I started "bee-watching"—trailing those buzzing critters in my yard so I could discover who they were and what they were doing. In Maine, as soon as the sun began dispelling the morning chill and lifting any lingering fog, I would take a stroll through my garden, while seagulls squawked and warblers sang their love songs. I would follow any tiny flying creature I saw, and I began to look more closely at my blossoms, often delighted to find tiny bees hidden inside. Any time I spotted a bee species I'd

• A tricolored bumble bee alights on a tansy flower.

never seen before, I was as excited as a baby meeting a dog for the first time. Conservation biologist Thor Hanson also had this sentiment, recalling in his book *Buzz: The Nature and Necessity of Bees*, "Just like walking with a toddler, pursuing something slowly and carefully sharpened the senses and created a whole new perspective." My garden in Maine was a flowery oasis in the midst of an evergreen forest, so of course it *teemed* with bees.

But I hadn't noticed them before.

Now I almost always see stripes zipping past me—pale yellow and dark black, yolk yellow and charcoal, arctic white and ebony, metallic green and black. My favorite is the big, burly, and loudly buzzing tricolored bumble bee, which is common in New England and gets its name from its furry rust, black, and yellow stripes.

And boy, are these tricolored bumble bees busy! There's no time to lose as they suck up nectar and grab legfuls of pollen, fertilizing flowers in the process. It was dizzying watching them flit from blossom to blossom, a flurry of activity from dawn to dusk while I loafed, drinking coffee, reading, watching the tide slowly come in and out. It's hard for humans to match the stamina of bumble bees: "Once I observed a single bee foraging for 16 hours, at which point I, not the bee, gave up," wrote researcher Lars Chittka in his book *The Mind of a Bee*.

Why are these bees always so busy? I realized I knew very little about a bumble bee's lifestyle, despite there being four times more bee than bird species north of Mexico, six times more kinds of bees than butterflies, and about ten times as many bee as mammal species. "Everyone knows that robins nest in trees, that bears hibernate, and that butterflies start out as caterpillars, but most people don't know where bees live, how they spend the winter, or what they eat," Joseph S. Wilson and Olivia Messinger Carril mused in *The Bees in Your Backyard*. Guilty as charged.

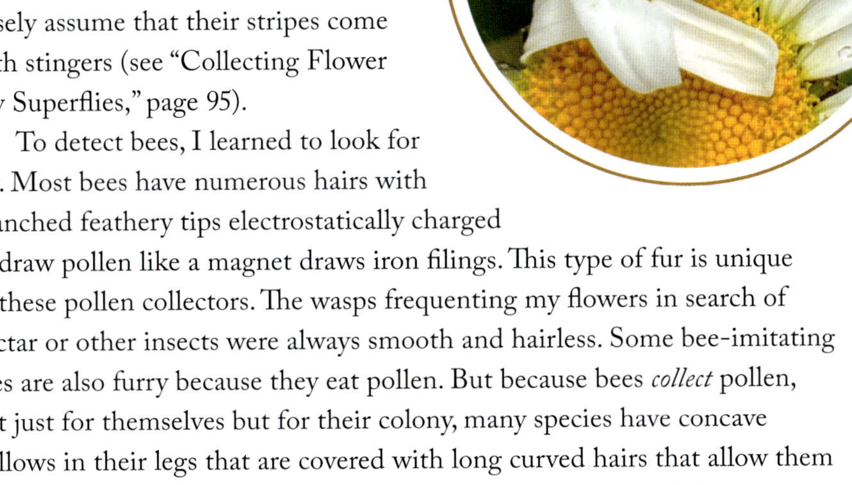

So I hit the books and the internet to learn more about the tricolored bumble bee and other bees I share my summers with. I soon learned that some of these "bees" were not bees at all, but wasps. Many others were flies mimicking bees to avoid predators, who might falsely assume that their stripes come with stingers (see "Collecting Flower Fly Superflies," page 95).

To detect bees, I learned to look for fur. Most bees have numerous hairs with branched feathery tips electrostatically charged to draw pollen like a magnet draws iron filings. This type of fur is unique to these pollen collectors. The wasps frequenting my flowers in search of nectar or other insects were always smooth and hairless. Some bee-imitating flies are also furry because they eat pollen. But because bees *collect* pollen, not just for themselves but for their colony, many species have concave hollows in their legs that are covered with long curved hairs that allow them to carry pollen back to their nests. It was these "pollen baskets" that gave the tricolored bumble bee I had first seen in my garden its garish orange pantaloons—a dead giveaway that it was truly a bee.

The dense fur on tricolored bumble bees also helps insulate them from the cold, enabling them to visit flowers on cool mornings earlier in the day and late into fall when all the other bees have vanished. As Hanson explained, if you glimpse something buzzing by and you're wearing "a wool hat, two layers of flannel, and a down vest, you're looking at a bumble bee."

• The legs of this tiny sweat bee pick up the yellow pollen from an oxeye daisy.

These bees can temporarily uncouple their wings from their flight muscles in order to contract them repetitively in a shiver that warms their bodies. This probably explained why I saw a tricolored bumble bee one frigid morning moving her body back and forth over a flower in a way that looked like she was having sex with a black-eyed Susan. Apparently, she was just doing her warm-up exercises for a busy day of dipping and diving, tugging and turning, and whatever else it might take to get pollen and nectar back to the little ones at home.

Eventually she warmed up, and I followed her as she zipped from black-eyed Susans to sneezeweeds and then skittered over goldenrods. She was a blur of a bee, a blizzard of movement, wings beating as fast as 200 times a second. There was no dawdling while she sucked nectar through her retractable strawlike tongue. A tongue that rooted the hyperactive bee to the flower for a mere moment so I could take pictures of her.

For a bee, there's no time to lose.

Once the tricolored bumble bee returns to her hive, she spits up her gathered nectar, mixes it with various secreted enzymes, and then air-dries it on her tongue to create honey. Unlike domestic honey bees, bumble bees produce only tiny amounts of honey. (Scientists have domesticated some bumble bees for farmers to use—not for their honey, but for their "buzz pollination" of hothouse tomatoes and similar crops. The pollen of these crops' flowers can only be shaken out of their pore-speckled anthers by the rapid wing-flapping vibration of bumble bee wings.) In fact, most bee species don't make honey at all. Before Europeans introduced the honey bee to North America, only bumble bees made honey here, but not enough for the resident humans to enjoy gathering and spreading it on their corn pone.

Social hive-and-honey-making bees are the exception rather than the rule. Most bee species live by themselves, each female building and provisioning her own nest alone, not within a colony. Bumble bees, though, are social, with colonies of 50 to 500 developing or adult bees—still much smaller than the 20,000 to 50,000 bees housed in a honey bee hive. Like most undomesticated bees, tricolored bumble bees nest inside cavities in the ground, such as in rodent tunnels or burrows, or in other natural small and shallow niches. (Thanks to being raised on Disney cartoons, we tend to visualize beehives as Japanese lanterns dangling from trees. But that's what paper wasp nests look like, not beehives.)

With the onset of cold weather, all tricolored bumble bees in the colony die except for several mated females. These bees are slated to become the new queens who will establish their own colonies in spring. Before winter arrives, they leave their nests and find a protected spot, such as a small hole in the ground, a mouse nest, or a pile of leaves, where they wait out the cold. It is only during this season that the tricolored bumble bee stills herself. No eating. No flying. No foraging. No laying eggs. She makes a natural antifreeze substance to protect herself from the cold. That and her slowed metabolism keep her alive until the growing season returns. (In contrast, honey bees remain active in their hives during winter and eat their stored honey.)

• Bees in their underground nest tending their immature siblings.

• The orange pollen baskets on this tricolored bumble bee's rear legs enable her to carry pollen to feed young back in the nest.

Once spring has sprung, the novice queen escapes her languid state, darts out of her hideout, and goes back to being busy. She buzzes around in search of an appropriate nesting site. Once she finds it, she forages intensely for several days, like a woman nearing the end of her pregnancy making multiple trips to the mall to gather supplies for herself and her baby. She frantically visits flowers, dining on pollen and bringing her pollen leftovers and nectar back to her nest. There she mixes the two together, adding a dab of secreted proteins to create "bee bread" that will feed her young. She also makes honey for herself and her progeny, and she stores it in a pot she creates by secreting beeswax from glands under her abdomen, working it into shape with special appendages on her legs.

Once she's filled her larder, the queen bee lays multiple eggs. You'd think she'd finally relax after such a flurry of activity getting everything ready for the next generation. But no, there's no laying low, slowing down, or gorging on sweets—there's too much work to be done! When she's not out foraging to feed herself, she must straddle the eggs and keep them warm with her pulsing abdomen contractions. This heat-generating shivering requires a lot of energy. During the monthlong incubation period, a queen might have to visit as many as 6,000 flowers to create enough honey to sustain herself. Once the eggs hatch, the queen will become a stay-at-home mom, tending her first batch of bee young, which start off as hungry blind white grubs.

She's able to stay at home because of her incredible ability to govern both the gender and the occupations of her progeny. To get help feeding her multiple large broods, the queen fertilizes most of her eggs, thereby ensuring that the first bees to hatch are females, and not those dastardly males who do no work for the nest and its young. The male slackers will hatch from unfertilized eggs the queen lays later in the season, along with female bees that are fed more than the workers so they can become future queens.

Male bees play only bit parts in the insect's life cycle. After spending a day or two feasting on bee bread or honey, they leave the nest, never to return, like Peter Pan's lost little boys. Some mornings I've spied these boy bees still sleeping on their blossom beds, encased in sparkling morning dew. During the day they forage just for themselves, and they live only a few weeks, dying once the cold ensues. In contrast, females spend most of their time foraging pollen and nectar for their siblings and queen back in the nest. The queen bee exudes pheromones (chemical communication signals) to ensure worker activity in the females and suppress their egg laying.

For the lucky one in seven bumble bee males who scores a female before his death, mating can last as long as an hour. To stave off competitors, the male bee remains attached to the female until the chemical plug he inserts along with his sperm dries. Despite mating being the all-encompassing highlight of a male's life, some queens continue to forage flowers during the mating process, undeterred by those annoying smaller drones latched on to them.

There's no time to lose.

Like the oldest children in a large family, the first batch of worker bees—all female—are soon put to work maintaining the nest and tending younger sisters. As they mature, they graduate to foraging for nectar and pollen. The organization of all these busy little bees is impressive, as is all the work they do, usually in exchange for a mere month or two's worth of life. It's exhausting and a little bit mind-boggling if you stop to think about it. There's no early retirement for tricolored bumble bees, and no time for leisure activities or creative endeavors other than those devoted to ensuring the success of the next generation. Even so, these busy teddy-bearish bumble bees are lucky—they aren't a declining or endangered species.

Other bee species may not be so fortunate.

Colony collapse disorder and other mysterious variables caused beekeepers in the United States to suddenly lose about half their hives in 2006 and 2007, creating panic in the agricultural community, which relies on these insects to pollinate crops. Europe and the Middle East also experienced a sudden, significant loss of their domestic bees. But there's both good news and bad news. The good news is that there are nearly 4,000 species of bees in North America and at least 20,000 worldwide. I had no idea! The world enlarged immensely when I learned that there are so many different kinds. Even though the population of domesticated honey bees is declining in the United States, native species could carry on the esteemed pollination tradition and should frequent my garden.

But now the bad news: In addition to the sudden 2006 loss of honey bee colonies (some of which are now recovering), 40 percent of all bee species worldwide are experiencing declining populations, according to a 2017 report. Similar research in the United States has found that more than half of native bee species are losing numbers, and about one-quarter are at risk of extinction. One study found populations of up to one-third of the forty-nine species of bumble bees in the United States declining, with three species vanishing from much of their range and another officially listed as endangered. These reports eerily echo Rachel Carson's prediction more than fifty years ago of the "silent spring" that could result if pesticides continued to decimate both bees and birds. I shuddered to think of a summer without the bees' familiar droning, let alone all the lost crops.

Several factors are fostering the decline in bee populations, including loss of habitat, competing invasive insects, and extensive single-crop farming. Widespread use of pesticides and newly introduced parasites and diseases also weaken or kill bees. Topping things off, climate change is making bees more susceptible to forest fires, floods, and other weather disasters, as well as creating a mismatch between when bees emerge from their winter dormancy and when the flowers they like to feed on are blooming.

Fortunately, there are things we can do in our own yards or urban spaces to help stave off bee losses, including providing good homes for them, replete with their life-sustaining flowers and overwintering sites. (See "If You Plant It, They Will Come," page 249.) We can also be more aware of the different kinds of bees around us so we can sound the alarm when they go missing. And we can be mindful that, as Helen Macdonald pointed out in her book *Vesper Flights*, "we are living in an exquisitely complicated world that is not all about us. It does not belong to us alone."

There's no time to lose.

FATHOMING BRILLIANT BEES

Family Apidae

LOCATION: On flowers in a garden, northeast Maine

DATE: May through October   TIME: Sunrise to sunset

I thought my babies were brilliant. Yes, I know, every parent thinks their children are above average, and so where are all those *below*-average children anyway? But every developmental step they took, no matter how tiny, made my babies brilliant compared to their helpless newborn selves. *Oh wow, he turned over! Amazing, in just eight months she learned how to crawl! Incredible, he can clap his hands!* Now, though, thanks to abundant new research on the mental capabilities of bees, I'm finding that my babies' brilliance pales compared to that of a newborn worker bee.

The first time this bee leaves her nest's dark womb, she encounters a panoply of sights and scents never experienced before. But she quickly adapts to her new environment. To get her fill of food for both herself and the

• An eastern carpenter bee feeding on wisteria.

brood back home, she gathers nectar and pollen from as many as a thousand flowers a day. Within a few weeks of hatching, a bee knows how to find these blossoms, has mastered their ins and outs so she can nab food from them, and knows the best time to forage for them. She can also navigate from nest to flowers and back, even when wind blows her off course. Such feats far surpass the ability to roll over at six months!

But wait, you say, that's not a fair comparison, because the bee is *born* with those abilities, whereas our children must *learn* everything. Yes, a bee is born with some of these abilities, like being able to fly. She's also especially suited for spotting flowers due to having eyesight fine-tuned to the range of flower colors and antennae apt at sensing flower fragrances. And a bee's built-in magnetic compass and innate ability to use the sun's position, or polarized light on cloudy days, helps direct her to and from blossoms.

But what's in bloom continually changes over the growing season, even varying by time of day. Because everything is in flux, a foraging bee's instinctual automatic responses won't suffice. She must learn to alter her plans based on information updated in real time in order to best gather her flower food. How does a bee accomplish all this learning in such a short period of time with no doting parent by her side to show her the ropes?

By being a quick study.

Extensive research reveals that bees become incredible learning machines from the moment they take their first flight. Newborn forager bees rapidly learn to link a flower's traits with its nectar and pollen output. Honey bees in one study were faster than fish and birds in grasping that a specific color is tied to a food reward. Human infants were the *slowest*.

But finding flowers with abundant nectar and pollen is only half the battle. After landing on a blossom, a worker bee must puzzle out how to tap its nectar. This process can be as complicated as opening a lock, and flowers vary considerably in their mechanics. Lars Chittka and his colleagues at

Queen Mary University of London set up a test using artificial flowers and found that bumble bees were remarkably fast at unlocking flower nectar. Impressively, the bees then applied what they learned to new flowers with similar structures, nabbing their nectar even faster. And they retained their memory of how to access the nectar in specific flowers for three weeks—the typical lifetime of a foraging bumble bee.

Bees also excel at waiting for a reward, a skill they need to keep themselves from drinking from flowers until the blossoms are filled to the fullest with nectar, which maximizes their foraging efficiency. (Different flowers have different nectar replenishment rates after their nectar has been sucked up by bees, other insects, or birds.) One study found that bees outperformed preschoolers in the bee equivalent of the famous marshmallow delayed gratification test. In this test, a child is presented with a marshmallow and told that if they wait a certain period of time without eating it, they'll get another one. Most young children fail this test, gobbling up the marshmallow immediately after the experimenter leaves the room. But Canadian scientists Michael Boivert and David Sherry successfully trained bumble bees to check themselves from sticking out their tongues for a sweet reward until after a specific interval of time passed. Bravo to the bees for their self-control!

A worker honcy or bumble bee may make a hundred trips each day to flowers as far as five miles from her nest. How does she find her way to and from home? To find out, Chittka outfitted some bumble bees with tiny backpacks that emit radar signals so he could monitor their movements. He found that a foraging bumble bee learns to recognize her home on her first flight by flying back and forth from the nest in different directions, each loop increasing in size over time. During this orientation flight, which is seen in many types of bees, the insect is probably learning the landmarks surrounding her nest. "These could be small stones, live or dead plants, bits of wood, or similar debris. She quickly creates a mental map of her home terrain," Stephen Buchmann explained in his book *What a Bee Knows*.

After that first orientation flight, she starts foraging in earnest, often following the same route to flowers for several days in a row. While foraging, she keys in to other landmarks to find her way, Chittka discovered. He and his colleagues placed identical bright yellow tents at regular intervals from a honey bee hive. Its inhabitants got used to visiting a feeding station halfway between the third and fourth tents. But when researchers removed some of these yellow tents along the bees' route, the bees overshot the distance to their hives, suggesting that the insects were counting the number of tents to find their hive. (Who knew bees could count?)

In other experiments, Chittka found that bees could also quickly learn to plot the most efficient route (out of hundreds of possible ones) for visiting all their flower foraging sites. "Bees had a consistent tendency to experiment—to try new ways to link the feeding stations, and if a novel route provided an improvement, to switch to that one," Chittka recounted in his book *The Mind of a Bee*.

Bees also excel at learning from each other. British researchers observing bumble bees noted that to access the nectar in tubular flowers, most of them took a cumbersome route, crawling deep inside the flowers to where the nectar pools. But a small group of innovators instead bit off the bottoms of the flowers (like I used to do to wild honeysuckle as a kid) to more quickly access nectar. The other bees watching them soon copied their behavior, taking the shortcut. Bumble bees also copy honey bees' flower choices.

Honey bees are even more refined in how they communicate with hive mates. They share directions to a good food resource with a carefully choreographed dance, their movements telegraphing the flight direction and distance from the hive based on the sun's position. Studies of honey bee populations in Sri Lanka found minor variations in these dances, much

• Different honey bee populations have different ways of communicating, suggesting that bees, like this Western honey bee, have varying cultural traditions.

like the different dialects of a language. This suggests that not only do bees learn from each other, but different bee populations may have different ways of doing things—cultural traditions based on what is "taught" in their communities. (Who knew bees had culture?)

Bee teaching was also seen by a UK-based research group led by Alice Bridges, who trained "tutor" bees to open a puzzle box in order to access a sucrose solution. The bees learned to open the boxes in one of two ways, and then the researchers put the boxes and the tutor bees in different colonies. They found that a tutor bee's specific way of opening the puzzle box quickly spread among the colony it was placed in. In colonies without a tutor, meanwhile, it took the resident bees longer to learn how to open the boxes.

The findings of all these studies suggest that the vast and complex repertoire of behaviors seen in many social insects are probably not innate, but rather *learned* behaviors that can vary depending on the cultural traditions of specific colonies. "The reason that we have often failed to see evidence of culture in some nonhuman animals may be that we are simply looking too late," Bridges and her colleagues wrote. Entomologist Jessica Ware, of the American Museum of Natural History, agrees. "Many of us consider ourselves and our fellow primates to be rather special . . . because we have culture and we can learn and we're social," she told NPR. But now, "it turns out even the bee also has culture, [and] that is an uncomfortable truth."

And here's an amusing aside: After training bumble bees to move a ball to the center of a circle to gain access to a sugar reward, researchers found the bumble bees rolling the balls even when they didn't get a sweet treat in return. Did the bees actually enjoy playing with the balls?

They tested this possibility with a careful design to rule out other reasons besides play that might have inspired the bees to interact with the balls when not rewarded. They found that not only did many of the bees play with small wooden balls about twice their size, comically rolling on top of them, but such bee play was more common in younger bees, just like it is in younger mammals and birds. (Experts claim such play behavior helps prepare them for future adult tasks.)

And boy bees spent more time playing with the balls than girl bees, presumably because they had more time on their hands (or should I say legs?). Remember that adult male bumble bees' sole purpose in life is to mate with a queen bee, but there were no queens in the experiment, only foraging female bees. So, faced with nothing to do, the boy bees played ball.

"This research provides a strong indication that insect minds are far more sophisticated than we might imagine," said Chittka, head of the lab that conducted the ball-rolling study. His colleague and primary investigator in the research, Samadi Galpayage, added that the study provides evidence for insects having feelings, including the pleasure gained by playing "like other larger fluffy, or not so fluffy, animals do. This sort of finding . . . will, hopefully, encourage us to respect and protect life on Earth ever more."

Okay, but I'm still wondering how bees' bitty brains let them do all that they do. It turns out that the unique architecture of a bee's brain allows it to accomplish a lot we previously assumed needs a bigger brain. The extensively fanned-out connections (synapses) between bees' brain cells (neurons) provide an exceptionally large capacity for remembering things, like where and when to find a flower with abundant nectar. (See "Investigating the Mind of an Insect," page 115, for more details.)

For me, a previously proud big-brained animal, learning about these brilliant bees has been quite humbling. I will reverentially think of the bee the next time I get lost trying to make my way back home or can't remember where I spotted that abundant patch of sweet-sour wild blueberries growing alongside the road. And, like the bee, I'm inspired to learn more—much more—about this incredible natural world in the limited time I have left on Earth.

PLAYING WELL WITH OTHERS

(Family Halictidae)

While taking pictures of a tricolored bumble bee feeding on a sneezeweed flower, I notice a gorgeous neon-green sweat bee less than half an inch long glinting like a sequin on a small petal nearby. There are close to 2,000 species of these tiny bees in North America, making them one of the most common bees on this continent—if you can see them. They range in color from shiny black to metallic shades of blue, copper, gold, and green, often striped with black or white.

These small bees are helping to answer a big question: How did social behavior, such as tending to siblings, evolve? That's because populations of the same species of sweat bees vary in how much they take care of their young. In colder northern regions, these bees' nests consist of a female who tends her progeny hatched from a single brood a year. But populations of the same species farther south show the same behavior as the tricolored bumble bee: The female egg-laying queen gets help rearing her young from her daughters, who tend younger siblings hatching in subsequent broods during the growing season.

Researchers discovered a genetic switch that turns on sibling-tending behavior in the southern population of sweat bees. But this switch is turned off in the northern population, where a shorter growing season usually precludes multiple broods.

And now for the clincher: Researchers deciphered that one of this genetic switch's key components affects the secretion of oxytocin in mice. Oxytocin is the so-called hug hormone because under certain conditions it prompts attachment, nurturing, and cooperation for a number of different mammals, including humans. Research also shows some links between a lack of oxytocin secretion and several disorders characterized by lower social function in people, including autism.

Maybe we're not so different from those insects after all.

• A bicolored striped sweat bee on a purple coneflower.

MARVELING AT DRAGONFLY SUPERHEROES

Autumn Meadowhawks (*Sympetrum vicinum*)
LOCATION: Garden in northeast Maine
DATE: November 7 TIME: 12:15 p.m.

It's early November in Maine, and low morning light gives a golden glow to the tawny grasses and silver puffs of goldenrod seedheads in the bay. Without flowers to feed on, there are no bees and few other insects flying around despite unusually warm temps in the sixties.

But for one striking exception: the autumn meadowhawk.

I'm delighted to see so many of these small carmine-colored dragonflies zipping around, sunlight shimmering on their translucent wings. I follow one until it stops to rest, and when I look more closely, I see not one autumn meadowhawk, but two, attached! One is lined up behind the other, and both

• Autumn meadowhawk dragonflies locked in tandem after mating.

"Overnight, the dragonfly nymph transitions from water to air, swimming to flight."

pairs of large chestnut-colored eyes stare at me, hexagonal glints in each. I knew mating dragonflies often fly together, the tip of one abdomen attached to that of the other. I've seen the wheel or heart shapes these pairs of dragonflies form while sperm transfers into the female's sperm storage sac used to fertilize her eggs. But these two autumn meadowhawks are lined up in tandem. Why?

I discover that after mating, the male autumn meadowhawk continues using his special claspers at the end of his body to grasp the female by the neck. Fitted together in this funny way, they fly to a body of water, where the male dips the female in repeatedly to lay her eggs. It sounds like doting coparenting, but this ensures no other dragonflies come along to mate with and deposit sperm in the female's sac.

Many male dragonflies have some behavior or specialized features to outcompete other males, given that often a female mates more than once before laying her eggs. In some dragonfly species, the male's penis equivalent sports lobes that expand during mating, packing down any sperm that preceded its own into the female's sperm pouch. In other species it is shaped like a bristle brush, the better to scrape out other males' sperm. Both remarkable designs boost the likelihood that the last one in scores a fertilization.

The males of autumn meadowhawks use the strategy of guarding their sperm by holding on to the females they mated long after copulation, sometimes until nightfall, creating what entomologist Gilbert Waldbauer called "living chastity belts." In his book *Insects Through the Seasons*, Waldbauer noted that the walking stick holds the insect record for this behavior; the male of this species remains attached to the female for nearly three months! Talk about being proprietary.

Some species of fruit flies have another surprising strategy to outcompete males in the fertilization game—supersized sperm. Although human men may obsess over the size of their penises, for these fruit flies it's the size of their sperm that seems to matter more. One species has sperm longer than two inches. Given the insect's tiny size, that's equivalent to a human sperm cell measuring the length of a tennis court, as Jonathan Balcombe pointed out in *Super Fly*. Much of its length lies in its tail, which forms tangles that block subsequent male sperm from fertilizing an egg. Who would have thought diminutive fruit flies, those flying specks circling our ripe bananas, are supersized in the sperm department? They hold the record for having the largest sperm cells of any known organism.

We don't see dragonflies for most of their lives because they spend at least a year or two as nymphs underwater in ponds and other bodies of water. The nymph is still recognizably dragonflylike, with its big bowl eyes. But it lacks wings and a long tail. It darts around using jet propulsion created by sucking up water and then rapidly squirting it out through its anus. The dragonfly nymph's rear

• The shed skin (exoskeleton) of a dragonfly nymph left behind on a pond reed after it transitioned to its flying adult form.

end has another important purpose—breathing—as its gills are lodged in its rectum. (Personally, I wouldn't want to breathe through my butt, but who am I to say this isn't a good strategy for the dragonfly nymph?) The nymph can quickly nab aquatic prey with its toothed retractable jaw like the shovel of a front-end loader. In a mere millisecond, this appendage jabs out and then retracts to scoop up insects or other prey. *Bam!*

Overnight, the dragonfly nymph transitions from water to air, swimming to flight. While clutching a reed, stick, or something else to keep it out of water, its back skin (exoskeleton) splits open and out comes a compressed dragonfly. This adult dragonfly gulps its first breaths of air while waiting for its wings and torso to expand and its new exoskeleton to harden. Imagine undergoing such a dramatic transformation—swimming one day and flying the next!

I marvel at the dramatic life cycle of dragonflies—one so unlike ours. I first came to appreciate dragonflies years ago when I was a damsel in distress and they came to my rescue. I was immersed in the middle of a marsh, my upper body marinated in sweat while muck painted my lower half. Cicadas rattled above, and heat and humidity warped the distant view. I struggled to ignore all bodily discomforts and focus on my master's thesis research, which required identifying the plants brushing my legs. Mosquitoes swarmed around me, feasting on my blood. Outnumbered, I couldn't stave off their itchy, annoying, and highly distracting bites with my futile swatting.

But then I heard the dim clatter of dozens of fluttering wings. Looking up, I saw not a bird or a plane, but a full-fledged squadron of dragonflies! Zipping in with large iridescent wings and cartoonish bulging eyes, they snatched one mosquito out of the air after another—*bam!*—like superheroes

• A twelve-spotted skimmer dragonfly. Because they have four independent wings, dragonflies can move in any direction in space, as well as hover like a helicopter.

in comic books. With their 95 percent kill rate, in the blink of an eye the dragonflies rid the marsh of all the mosquitoes that had been plaguing me for the past hour, saving the day, as they say. From that moment on, dragonflies became my heroes for life, my best friends forever. Obsessing over them, I collected anything with their images, from shirts to mugs to earrings. Now I collect dragonfly facts while trying to understand how they so quickly cleared the marsh of mosquitoes.

The more I learn about dragonflies, the more they astonish me. One of the first winged insects to evolve, dragonflies watched the dinosaurs sink into the ground, destined to become fossils (or fossil fuel), while they kept going. Like many animals when the world was young, dragonflies were bigger—much bigger. Back then, their wings spanned more than two *feet*.

Evolution perfected dragonflies' eyesight over more than 300,000 years, so they are perfectly equipped to snatch fast-flying prey. Their enormous eyes, comprising most of their heads, have 24,000 facets, enabling them to see in every direction except behind. These eyes capture 200 images a second. (Human eyes capture a paltry 60.) This fast-snapping-camera vision lets dragonflies experience time differently than us. They see life in slower motion, so they can react more quickly—in as little as 30 milliseconds—to prey and other things we never notice.

Because they have four independent wings, dragonflies can also move in any direction in space—sideways, forward, backward, upward, and downward. And, like a helicopter, they can hover in a single spot in the air for a minute or more. With these aerial feats, dragonflies can ambush unsuspecting flying insects from any direction. Their legs gather and point upward to create a death basket that scoops up prey in flight. With all these superpowers, no wonder they were able to nab all those mosquitoes swarming me in the marsh!

Not only are they agile, but dragonflies are fast—they can zip about at more than thirty-five miles per hour. That speed is due in part to the intricate and prominent veining of their wings, which not only underscores their iridescent beauty but stiffens and strengthens them so they can pull more distance with each wingbeat. Those veins are also equipped with more than 3,000 sensors for airflow and wing strain that help dragonflies fine-tune their flight—hence their precision in nabbing prey. Inspired by the flying finesse of dragonflies, engineers are currently modeling new designs for aircraft, cars, and wind turbines after them.

Dragonflies need strong wings because some species in North America migrate over a thousand miles south each fall, along with the birds. Between India and East Africa, multiple generations of globe skimmer dragonflies clock more than 10,000 miles each year. Here in the northeastern United States, where I live, green darners and various species of skimmer dragonflies fly all the way down to the southern states or even farther each fall. Mass movements of autumn meadowhawks also occur in the Northeast in fall, perhaps explaining why I saw so many in November, but scientists have yet to document any major southward migration for the species.

Dragonflies' migratory journeys are terribly tiring. They often have to rest motionless during brief stopovers or stay in one spot until their muscles warm up for flying. Many don't survive. They litter the ground with their chitinous carcasses when days grow cold. One fall day, while out walking with a friend, I spotted a large immobile dragonfly on the road that I assumed was dead. Wanting to bring it back to adorn my desk, I picked it up and carefully cupped it in my hand during our walk. But ten minutes later it jolted me with its movement! Not dead, it was merely resting or warming up.

Opening my palm, I let it flutter away, wishing it a safe journey.

• The large eyes on this mosaic darner dragonfly capture more than three times as many images a second than the human eye.

CATCHING THE *PREYING* MANTIS

Praying Mantises (Family Mantidae)
LOCATION: Edge of a trash can in a Philadelphia park
DATE: August 30 TIME: 2:17 p.m.

It robotically swiveled its head to meet my gaze with big bug eyes. There was something Tinkertoy-ish about this giant insect, as if a child put together its body parts. A triangle for a face, which can turn all the way to the right or left on its long neck. A long spindly torso with limbs awkwardly angled. Front folded legs, serrated and spiked. Elongated antennae.

This praying mantis, as long as the palm of my hand, perched on the edge of a Philadelphia park's aromatic trash can . . . waiting. Waiting for unsuspecting flies to come its way. Waiting for something flitting. Waiting for dinner. To better spot its prey, it rocked back and forth repetitively like an Orthodox Jew davening, its folded forelimbs held close to its chest

• Praying mantises, like this one, ambush their prey. They wait until unsuspecting insects land near them, then snap their front legs shut, impaling prey with their spines.

resembling someone in prayer—hence its name. But flies circling the trash wouldn't *have* a prayer once it clasped them in its piercing forelimbs.

This was a *preying* mantis.

Mantises, which can grow to up to eight inches in length, are such good hunters that they can catch and kill small rodents, lizards, frogs, and even birds, along with any bugs coming their way. Ambushers, they lie in wait, aided by their astonishing ability to camouflage and blend in with their surroundings. The one my son and I found wore a skirt of closed chartreuse wings perfectly mimicking spring green leaves—something designed to hide it among vegetation, not on a trash can.

Some mantis species in the African and Australian bush can molt to black and visually disappear on scorched vegetation in fire-ravaged areas during the dry season. Those that roam the bare ground have flatter shapes to eliminate any shadows that might reveal their presence to prey. Particularly fanciful are the flower mantises, including orchid mantises, that take on both the structure and color of the flowers on which they lie in wait. But these are no demure wallflowers. When unsuspecting insects come close, their pink frilly "petals" snap shut—those frills are actually spines that impale the insects before they can partake of any flower food.

Let's not forget—they are *preying* mantises.

"Can we bring it home?" my ten-year-old son asked, enamored with this seemingly alien creature. "Sure, why not?" I responded, thinking praying mantises might be good for my garden. (I was mistaken. It turns out they'll eat any insect that comes along, pest or pollinator.) So we brought the praying mantis back with us and set her among my flowers in front of the house. We spotted her for several days afterward, but then she disappeared, or maybe we stopped noticing her, as often happens with insects. One day she was all my son could talk about, and then the next day he had moved up to the vertebrate level, pestering us for a pet lizard.

But the praying mantis had *not* disappeared.

The following spring, while digging in my garden, I spotted several miniature praying mantises, each as long as the tip of my pinky, slowly plying their way through the vegetation. For these tiny creatures, a daffodil leaf can be like a never-ending green highway in the sky, and a traipse to the nearby zinnias a safari. They were so cute with their tiny triangle faces—and the spitting image of their mother, minus her wings. Apparently, in the fall she had laid eggs that overwintered and sprouted into baby praying mantises once the weather warmed. As summer progressed, I saw larger versions in my garden, which all looked identical to their younger selves, although they eventually added wings.

We humans can expand our body parts, like breasts or testicles, as we age because they are hung on an interior skeleton and we have skin that can stretch and grow around them. But praying mantises and other insects have a rigid exterior skeleton that can't accommodate expansion and change. To grow, they have to repeatedly and regularly shed their exoskeletons.

Sometimes that molting is accompanied by morphing into a completely different being. For example, a caterpillar sheds its skin and rearranges itself while hidden within a cocoon in its pupal stage, then emerges as a moth. More than three-quarters of insects undergo this type of stunning complete metamorphosis so they can tap different food resources in different stages of their development. Leaves are the first foods to arrive for caterpillars to dine on. These are followed by the sweet nectar and pollen of flowers the later-developing moths can consume. But other insects, like our praying mantises, stubbornly stay nearly the same shape as each bigger version of themselves, called an instar, busts through. Many go through five to ten instars before reaching their final adult form with wings.

Interior skeletons may offer more flexibility, but exoskeletons can be incredibly strong and durable, offering more protection for soft fleshy parts. Consider the aptly named diabolical ironclad beetle, which is virtually impossible to squish, even though it's only an inch long. On YouTube, you can find a video of a Toyota sedan running over a specimen of this beetle twice, yet it miraculously emerges relatively unscathed. Like a tick you can't squish, only more so, the diabolical ironclad beetle has an exoskeleton layered with repeating structures, overlapped to disperse forces that beset it, making it the armored knight of the insect world. It can withstand nearly 40,000 times its body weight—the equivalent of a person resisting the crushing force of about 1,800 Toyota sedans. Insects like this beetle are inspiring scientists' work on designing stronger buildings and uncrushable vehicles.

For their part, praying mantises have been inspiring cartoonists to capitalize on the cannibalistic tendencies of the female, who sometimes swivels her head to bite off the head of a mating male mounting her back. One cartoon by Danny Shanahan depicts two praying mantises, one saying to the other, which is missing its head, "You slept with her, didn't you?" Astonishingly, having its head bitten off during sex doesn't stop the male from inseminating the female, but merely turns sex into a mindless venture. A decapitated male can carry on mating with the female for several hours while she licks off the last bits of his head from her forelegs like a cat grooming itself. She eventually consumes the rest of her mate.

But here's the truth: Such cannibalization tends to happen only when the insects face limited food resources or strange captive environments. As entomologist Gilbert Waldbauer pointed out in *Insects Through the*

• Can you find the well-camouflaged praying mantis hidden inside this orchid? (Hint: Look for its two antennae poking out of its triangular head.)

Seasons, the sacrificial male "becomes a meal that will help [the female] to survive until she has laid the eggs that will become his sons and daughters," thereby carrying on his lineage. But in labs where mantises are fed as much as they want, such cannibalism doesn't tend to occur. At the University of California-Santa Cruz, neurophysiologist W. Jackson Davis observed thirty pairings of praying mantises and found that in none of them did females devour the males. In fact, he saw the males performing elaborate courtship dances, such as swinging their antennae forward and backward and repetitively curling up their abdomens over their backs. Or they strutted sideways, doing a rhythmic two-step with their rear legs. Scientists sometimes even observed female mantises stroking the forelegs of their male consorts. "If [they] were human, you'd have to call it foreplay," Davis remarked, as quoted in a *Washington Post* article by Boyce Rensberger.

In truth, like many other animals, praying mantises are more likely to eat their siblings than their mates. Despite their infrequent coital cannibalism, which is estimated to occur less than a quarter of the time in the wild, the female praying mantis is still seen as a femme fatale. It prompted surrealist artists such as Salvador Dalí and Alberto Giacometti to obsess over the insect as a symbol of a preying woman, with the sex act one of self-destruction and death for the male, rather than procreation for the species. These artists, some of whom kept praying mantises as pets and invited people to view the insects' matings, also frequently used toothed vaginas in their artwork to symbolize the castrating nature of women. "As a symbol of male primordial fears and needs, the mantis image in Surrealist art is mostly negative. . . . The Surrealists' fascination with the mantis is an indication of their attitude toward woman, which was ambivalent, if not misogynistic," noted art historian Ruth Markus in the 2000 Spring/Summer issue of *Woman's Art Journal*.

"Like many other animals, praying mantises are more likely to eat their siblings than their mates."

What made these men fear women and their own sexuality? Was it their lack of control over their own desires? It reminds me of a boyfriend I once had who, after two-timing me, apologized by arguing that a man's brain was located in his loins, like the praying mantis who keeps squirting sperm even after his head has been bitten off. Is that why the sexually cannibalistic praying mantis is so often featured in popular media, even though she is more the exception than the rule?

Perhaps we need to view this exception in a more favorable light. After all, for most species in the animal kingdom, once he squirts his sperm, the male's job is over, as far as procreation is concerned. Most males don't sit on eggs or protect those eggs from predators. No males experience a fetal takeover of their body, contorting it in every which way and pillaging its bones and blood for nutrients. No males (with the exception of fruit bats) breastfeed. So maybe it's okay that a male praying mantis sometimes sacrifices himself for the sake of his progeny. And it's no surprise that his starving mate is sometimes compelled to bite off his head to get the resources she needs to carry on both his and her lineage. After all, it's in her nature, a mindset shaped by the pitiless force of evolution.

Let's not forget—she is a *preying* mantis.

• A newborn praying mantis sits on the tip of my finger.

FINDING TINY MIRACLES HIDDEN IN PLAIN SIGHT

Swallowtail Butterflies (*Papilio polyxenes*), Atlantis Fritillaries (*Speyeria atlantis*), Great Spangled Fritillaries (*Speyeria cybele*), Monarch Butterflies (*Danaus plexippus*)

All you need to witness a miracle is some parsley.

My townhouse, in the heart of Philly, has a miniscule, mostly flagstone courtyard. One day in late May while reading on this patio, I spotted an elegant black swallowtail butterfly wafting by with its yellow, orange, and sapphire spots on three-inch-wide black wings like jewels encased in a black velvet box. Captivated by her beauty, I watched as she kept butting up against the bit of parsley growing in a small patch of exposed soil. *What is she doing?* I wondered.

After she left, I spotted tiny yellow dots on a few parsley leaves—her eggs! I checked them every day for hatching caterpillars, and sure enough, a week or so later I saw a tiny, spiny, mostly black one with a big splat of

• A swallowtail butterfly perched on a goldenrod stalk.

white on its middle, designed to mimic bird poop (see "Unmasking Insects Blending In," page 89). This early rendition (instar) of the caterpillar looked nothing like the much bigger version of itself that I saw several weeks later, still munching away at the parsley. It had grown fat and now had vibrant green-and-black stripes with bright yellow spots—a flamboyant clown version in comparison to its more serious and subdued younger self. (And who wouldn't be subdued if they looked like bird poop?) The older caterpillar's contrasting colors helped it blend in with the foliage in dappled sunlight, like a zebra's stripes.

But there's no hiding from me. When I plucked the caterpillar from the parsley and held it in my hand, it surprised me by rearing up on its pudgy legs. Then out from its head popped what looked like a bright orange forked snake's tongue. *What's up with* that*?* I wondered. Later, I learned that this appendage spews out a chemical repellent to deter other insects, like ants, from eating the caterpillar. Its snakelike appearance may also serve to frighten away any birds hoping to eat it.

Black swallowtail caterpillars don't eat just parsley; they dine on other plants in the carrot family, including fennel and Queen Anne's lace. Other butterfly caterpillars are choosier. The monarch caterpillar will eat only milkweed leaves, while violets are the only food for the young'uns of the orange-and-black-checkered Atlantis and great spangled fritillary butterflies we'll explore later. But like human children, once these picky-eater caterpillars become adult butterflies, they expand their diet. Most butterflies will unfurl long, spiraled, straw-like tongues to suck up flower nectar, feeding on a much wider variety of plants than they did as caterpillars. They no longer munch on leaves because they've lost their chewing mouthparts. Some moths and butterflies don't even eat at all. Instead, they rely on their caterpillars' feeding to give them all the calories they need for the few days they are alive.

• The life cycle of an insect as painted by Maria Sibylla Merian in 1705.

I wanted to watch the black swallowtail caterpillar morph into a chrysalis and then emerge as a butterfly. To do so, I put it in a jar with some parsley sprigs and capped the jar with a paper towel, through which I'd poked some holes for aeration. I felt like I was channeling seventeenth-century artist and naturalist Maria Sibylla Merian, who was one of the first to document the life cycles of butterflies and moths. Prior to her scientific observations, people didn't link the caterpillar to its pupal cocoon or chrysalis, let alone to its adult moth or butterfly form. Instead, all these stages of life for a single insect were seen as distinct, unconnected beings.

That changed after Merian. Enchanted by the beauty of insects and curious about their lives, she spent most of her own life closely observing them in nature, raising hundreds of caterpillars in her home so she could sketch and paint them and the plants they lived on. She published three volumes of her illustrations of the metamorphoses of butterflies and other insects, fleshing out their life cycles by depicting each stage on its host. Her illustrations and written descriptions were so detailed that Carl Linnaeus, creator of the scientific system of naming organisms still used today, relied on them to define some species for which he lacked specimens.

Some people thought Merian's discoveries of insect metamorphoses went contrary to the Bible. If God created blueprints for every type of animal on Earth, it was sacrilegious to think an animal could dramatically change its own form. But Merian countered in the introduction to one of her caterpillar books that the insects and their metamorphoses reflect God's glory and that she wished to put "such Godly miracles in a little book."

As for the miracle I was hoping to witness, less than a week after capturing the black swallowtail caterpillar, I noticed it scrunched up and immobile on a parsley stem. A day later I was thrilled to discover its mottled green, yellow, and brown chrysalis. It resembled a bud or curled-up leaf, but looking closer, you could see the artful construction. This chrysalis leaned out from the stem, its bottom end anchored to it with a Velcrolike hook. Like a safety harness for a tree climber, a thin loop of silk was wrapped around its upper end, which extended out from the stem at an angle. This suspended position gives the butterfly the room it needs to unfold its large wings when it eventually crawls out of the chrysalis. The well-camouflaged chrysalis hides the caterpillar while it transitions into a butterfly.

Once the butterfly emerges, its sole purpose is to mate and lay eggs. This life-renewing effort relies on its newly grown wings, which it uses to seek out mates. But the manner in which a caterpillar transforms into a butterfly or moth inspires complete awe.

• A black swallowtail butterfly chrysalis suspended on a stem of parsley.

Wings start to develop in the caterpillar before it even creates its chrysalis. Portions of its head remain while it pupates and expands to form various parts of the butterfly. But during the two weeks or so that the swallowtail caterpillar is encased in its chrysalis, food-digesting enzymes will dissolve nearly all the rest of its tissues. The exception is small groupings of cells called imaginal disks. Under the influence of various hormones, these stem-cell clusters rapidly expand in number to form the butterfly's anatomy.

This transformation is still miraculous to me, despite science starting to be able to explain the genes that commandeer it and the proteins they create. Think about it: The caterpillar has to develop wings, compound eyes, long spindly legs, sex organs, a spiraled straw tongue, and all the muscles and neural wiring needed to support flight and new eating and mating behaviors—all in just a few weeks. Plus it has to craft delicate, colored scales to cover its wings in the pattern unique to its species. Astonishing, right? I wished I could have peeked inside the chrysalis and seen its wondrous molecular machinations in action, like some researchers have.

But I didn't even get to see the butterfly emerge. Instead, one morning I was surprised to see the swallowtail with its wings fully extended and nearly touching the edges of the jar, having left behind the chrysalis shell. Like an insect version of *The Ugly Duckling*, the bird-poop mimic I had plucked from my parsley had grown into a startlingly beautiful butterfly, and I felt like a proud parent. I brought the jar outside, opened the lid so the butterfly could escape, and wished it well in its efforts to establish a brood of its own. The second-generation caterpillars hatching from its eggs would likely overwinter in their cozy chrysalises, although if we were in some warmer locale, they might emerge as butterflies and create a third brood during the growing season. Females always lay their eggs on their caterpillars' host plants so their offspring will have a full larder right at their fingertips—oops, I mean feet—when they hatch.

After seeing the entire swallowtail butterfly's life cycle, I was eager to witness another miraculous transformation and started reading up on fritillary butterflies, herds of which flock to the lilac-colored spikes of the veronica flowers that bloom in my Maine garden in summer. With so many descending into my yard, surely I'd have a chance to see some of their caterpillars and could then watch them become butterflies?

There are thirty species of fritillaries, but the ones we see up in Maine are generally the Atlantis (*Speyeria atlantis*) and the great spangled (*Speyeria cybele*). The common name *fritillary* comes from the Latin word for chessboard and refers to the markings on these butterflies' wings. Both species of fritillary are light orange and checkered with black spots, dashes, and chevrons. The great spangled is a little larger than the Atlantis—though that's hard to gauge unless they're side by side. Then I discovered a more reliable way to tell them apart in photos—by their eye color! Yes, those multifaceted eyes come in different shades for different species and are not the black beady types you find in many insects. The Atlantis fritillary has grayish-blue or greenish-blue eyes, while the great spangled fritillary has amber or yellow-green eyes. So the next time you want to get the name of a fritillary, look into its eyes. (Speaking of names, "great spangled" refers to the silver spots on the species' lower set of wings when they are closed.)

• A fritillary dines on butterfly milkweed in my garden.

The summer I liberated the swallowtail butterfly I had raised in captivity in Philadelphia, the veronica flowering stalks grew abundantly in my garden in Maine and towered over my head because the deer hadn't browsed on them and we had plenty of rainfall. How incredible to behold dozens of fritillaries lining up one above another like a regiment to feed on the purple flower spires! If I got too close, they chased each other up into the sky in a swirl of color, a blizzard of butterflies.

But no caterpillars—at least none I could find.

Caterpillars can be quite elusive, hiding under plants or plant debris on the ground and active only at night. The caterpillars of most fritillaries, including both the Atlantis and the great spangled, eat only the leaves of violets—something I'm mindful of when gardening, making sure not to weed out those precious wild plants. You'd think, then, that fritillary butterflies would be certain to lay their eggs on the leaves of violets so their young'uns would have no trouble finding them once they hatched. But when most fritillary butterflies mate in midsummer, the violets have already died back. So fritillaries opt to lay thousands of eggs instead of the typical hundreds that other butterflies lay. They lay them in shady, weedy spots where violets are likely to grow. It's possible that some fritillaries have a nose—or, more accurately, antennae—for the roots of violets and lay their eggs where they smell them. Despite that, many fritillary caterpillars will die because they can't find their food source. But given their large numbers, this egg-laying behavior has proven sustainable for these species.

Prior to laying their eggs in late summer, female fritillaries may briefly hibernate under the shelter of vegetation to escape the heat of the season. Their caterpillars also spend a lot of time in dormancy; although the Atlantis fritillary and great spangled fritillary caterpillars hatch two to three weeks after the eggs are laid, they merely drink water and then go to sleep for seven to eight months, with only some making it through the winter. Once spring comes, surviving caterpillars dine on whatever young tender violet leaves they can find. You'd think they'd have evolved an easier lifestyle, but who am I to argue with Mother Nature?

Global climate change might harm fritillaries if it causes violet leaves to come out earlier and die before the fritillary caterpillars wake from their dormancy. Researchers, with the help of citizen scientists, like those who document what they see on iNaturalist, are currently trying to assess if such a mismatch is occurring. Hopefully, the Atlantis fritillary won't disappear like its mythical namesake.

Instead of pupating inside a chrysalis, the Atlantis fritillary uses its silk to sew together a tent made of leaves and undergoes its transformation into a butterfly while hidden inside this enclosure. The great spangled fritillary creates a sturdier chestnut-colored chrysalis and makes a thin line of silk to suspend it from a branch, piece of bark, or other low-lying object near the ground. I've yet to find a fritillary chrysalis but intend to spend more time combing the ground in the hope of spotting one.

Unlike their younger caterpillar selves, great spangled and Atlantis fritillary butterflies aren't picky about what they eat. They will sip nectar from just about any sun-loving flower, although they tend to prefer long, tubular ones.

Some people confuse fritillary butterflies with monarch butterflies (*Danaus plexippus*) because both have orange wings with black markings. But the wings of the latter are a darker orange and resemble stained glass more than a checkerboard. Monarchs were the first to give me a glimpse of miraculous metamorphoses. By August I usually see their chubby caterpillars, striped in black, lemon, and white, chowing down on my garden's milkweed. Sightings of monarch caterpillars and butterflies bring me special joy because this species has been scarce for the past decade, due to its rapidly shrinking habitat and expanding milkweed-killing herbicide use. The International

Union for Conservation of Nature considers the monarch a vulnerable species, and the US Fish and Wildlife Service has proposed listing it as threatened.

In response to dramatically declining monarch populations over the past few decades, many gardeners have taken to growing milkweed in their yards. Other monarch-lovers are raising monarch eggs and caterpillars indoors, in captivity, to protect them from predation, then setting the butterflies free once they emerge. In the wild, only one in ten monarch eggs result in butterflies; these people hope to improve those odds.

But according to the Xerces Society for Invertebrate Conservation, this practice, when done on a large scale, may hinder more than help the species, for a number of reasons. Raising monarchs in captivity creates denser conditions than occurs in the wild, making them more susceptible to disease and parasites. Captivity might also make them more likely to starve, as more caterpillars and butterflies may survive than there are resources to feed them.

Also, in the wild, natural selection ensures that only the fittest individuals survive. This makes them stronger and more resilient as a species, unlike captive breeding. Companies raise genetically limited monarchs so a flurry of butterflies can be released at weddings and other events. But this can dilute the diversity that the butterfly population needs to survive as a species. One study that recovered tagged monarchs found that captive-bred butterflies had lower migration success compared to wild monarchs. The Xerces Society

• A monarch butterfly caterpillar crawls on my butterfly milkweed plant.

recommends rearing no more than ten monarchs per year in captivity and raising them in individual containers that are disinfected before being used again.

I have raised a few monarch caterpillars indoors in the past so my daughter and I could watch them transition into smooth, shiny, and quite elegant chartreuse chrysalises—jade amulets with striking necklaces of gold, edged with black, encircling their conical tops. It was wondrous to watch the butterflies emerge compressed and crinkled and then take their first flight once their wings expanded and dried. But now I let nature take its course when I spot monarch caterpillars on my milkweed. Rather than raise them indoors, I instead feel blessed by every monarch butterfly I see sucking nectar from my flowers or sailing down the coast, and I'm inspired by their arduous long flights (see the "Miraculous Migrations" box, page 48). I feel similar joy and amazement when I see a flurry of fritillaries flocking to my veronica, or when I spot a swallowtail butterfly drinking upside down from my lilacs. These marvels of nature reward me with their hard-earned beauty, and their chrysalises remind me of the astonishing transformations going on around me all summer long—each a tiny miracle hidden in plain sight.

- Above: A monarch butterfly perched on a verbena plant in my garden.
- Right: A bee shares a stem with a monarch chrysalis.

MIRACULOUS MIGRATIONS

Monarch butterflies east of the Rocky Mountains are famous for their epic migrations, which they make each year from their wintering grounds in Mexico to their summering grounds in the upper stretches of North America. As I wrote in my book *More Than Meets the Eye*:

> It takes three to four generations of monarchs to make this journey of more than two thousand miles. Each butterfly lives about two months and travels only part of the way before laying eggs on butterfly weed and other milkweed species. These eggs enable the next generation of butterflies to carry on the trek north. The final generation that makes the return flight to Mexico can live as long as nine months. . . . Still a mystery is how the monarch butterfly, over a span of so many generations, knows to return each year to the same forested slope in Mexico to winter—the equivalent of long-dead great-grandparents dictating where their great-grandchildren spend their winters.

Amazing, right? But also amazing is how this migration was discovered in 1975, prior to the miniaturized transmitters used today to track butterflies, and before organized citizen-science endeavors like iNaturalist. Starting in the 1950s, to uncover the mystery of where the monarchs seen in North America went every year, husband-and-wife researchers Fred and Norah Urquhart of the University of Toronto deployed a network of hundreds of volunteers, including school groups and naturalists. These volunteers put stickers on the wings of thousands of monarch butterflies that asked those finding the labeled butterflies to return them to the biology department at the University of Toronto. This endeavor revealed that the butterflies migrated from south to north in the spring and from north to south in the fall. But it wasn't until the butterflies were spotted in a forest in Mexico that it became apparent just how far they could travel.

Another miracle hidden in plain sight.

• Each fall, monarchs east of the Rocky Mountains migrate thousands of miles to their wintering grounds, seen here, in Mexico.

GLIMPSING AN ETHEREAL VISITOR

Luna Moths (*Actias luna*)

LOCATION: Brushing against a windowpane in northeast Maine

DATE: June 28 TIME: 9:08 p.m.

The apparition flutters in front of my dark window—a giant pale green moth, like a delicate lady's handkerchief, with striking crescent moons embroidered on each of its almost translucent wings. These wings taper into something long and trailing, reminding me of the amorphous limbs of a revenant gliding by. Brushing against my windowpane, the otherworldly luna moth glows in the blackness, akin to its namesake, jolting me from my somnolence.

An awe-inspiring wake-up call.

The next morning I spy the moth resting on a tree and further admire its graceful form and five-inch wingspan—it's one of the largest moths in

• The luna moth, with its five-inch wingspan, is one of the largest moths in North America.

North America. But I'm puzzled by its long, twisted tails, which resemble a ballet dancer's legs turned out on point. Why have them? To make the moth more aerodynamic? To appeal to potential mates (or admiring humans)? I comb the internet and find none of these answers correct. The right answer surprises me: The tails scatter the sonar signals that bats emit to find their prey, leading the bats to miss the moth or attack its tails rather than its more essential body.

This trick appears to work. The next few mornings, I find the luna moth still with us, resting on a spruce tree. Then she vanishes as mysteriously as she had appeared; because, alas, these stunning sylphs live no more than a week. A week in which to mate and lay eggs for the next generation. A week during which they fast, lacking a digestive tract and functional mouth. A mere week to spread their gossamer wings and explore our world before departing into a truly ghostly realm, leaving us with the transformational memory of their diaphanous splendor.

To ensure the next generation's creation, female luna moths exude a distinctive mixture of chemical compounds comprising its sex pheromone. Alighting on the large, feathery antennae of the males, this pheromone sets them on fire with desire, drawing them from miles away when the wind blows right. A different moth led nineteenth-century French naturalist Jean-Henri Fabre to first postulate the existence of such amorous compounds one moonlit May night. Pleased to have had a great peacock moth emerge from a cocoon in his laboratory-study earlier that day, he put her under a wire-gauze bell jar on a table. Around nine o'clock that evening, Fabre's pleasure turned to enchantment as dozens of wan gray giant peacock moths, each as large as a hand, with eerie eyespots on their wings and antennae plumes on their heads, floated in through the open doors and windows of the house. "Coming from every direction and apprised I know not how," Fabre wrote in his *Life of the Caterpillar* (first published in English in 1916), "here are

• I discovered a chubby luna moth caterpillar while biking in Maine.

forty lovers eager to pay their respects to the marriageable bride born that morning amid the mysteries of my study."

Over the following week Fabre caught more than 150 of these moth suitors. No matter where in the house he moved his newly emerged moth, the male moths found her. What drew them to her each night like zombies from the grave? Over the next several years Fabre carried out painstaking experiments to learn the moth's secret. Eventually he concluded that even though no human nose could detect it, the female moth must release an odor powerfully attractive to the opposite sex of her species.

It wasn't until the twentieth century that scientists successfully identified the mysterious sex pheromones wafting in the "odors" emitted by female moths. Such enticing compounds enable them to find a mate, which otherwise could be challenging considering how immense their environment is compared to how small they are. In such a scenario, chemical signals have many advantages over visual or acoustic cues: They can be highly effective in the extremely low doses a tiny creature is likely to produce, as well as over the extremely great distances often needed to find other members of the tribe. Chemical signals also persist, can be detected in the dark when many insects are active, and can pass over enormous physical obstacles obscuring a tiny animal's sightline.

Insect pheromones are astoundingly, if not quite magically, potent. One scientist observed a single caged female pine sawfly attract more than 11,000 males from the field in less than five days. Other researchers documented a male promethea moth flying at least nine miles in three days to pheromone-releasing bait. But here's my favorite pheromone story: Another scientist amusingly reported that he persistently drew male spongy moths to him after inadvertently spilling a bit of the female spongy moth sex pheromone onto his skin and clothing. What a spectacle—I imagine him akin to Mauricio Babilonia, the character from Gabriel Garcia Márquez's *One Hundred Years of Solitude* who was always trailed by a cloud of yellow butterflies.

Female moths and butterflies aren't the only ones releasing sex pheromones. Some male moths, flies, butterflies, cockroaches, and bees emit aphrodisiac pheromones to get females in the mood. And some female insects produce *anti*-aphrodisiac pheromones to discourage suitor advances. These include a noxious substance a female ground beetle squirts at males who just don't get it and persist in courting her when she's already been mated, like those guys in bars who won't leave happily taken gals alone. There's even an insect anti-aphrodisiac that acts like a chastity belt. Made by a male tropical butterfly, it's a plug that not only blocks the female's genital opening but has a smell so offensive it repulses other suitors.

But pheromones govern more than just the sex lives of insects—they serve as an all-purpose, species-specific chemical language for these small beings. Researchers have cracked thousands of insects' pheromone communication codes and found that these compounds guide much of insect development and behavior. For example, when laying eggs, some flies, moths, and beetles use pheromones to repel insects of the same and competing species, ensuring their hatchlings are able to access limited resources. Queen bees emit a pheromone that suppresses the sexual development of worker bees so they care for her progeny and don't create their own. Ants use pheromones to direct nestmates to a food source, which explains the trails they make at picnics or in your kitchen. "Leafcutter ants are so sensitive to their trail pheromone that a milligram is enough to lay a path around the planet three times over," Ed Yong noted in his book *An Immense World*. Aphids give off alarm pheromones to urge neighboring aphids to flee from nearby predators. Honey bees use them to recruit nestmates to sting and pursue intruders. One species of ant produces some thirty alarm pheromones that, depending on the composition and concentration of the mixture, send signals that fine-tune the behavior of worker ants. The list goes on and on.

Akin to the way humans can combine a limited number of words into sentences to convey multiple meanings, insects can mix different pheromones in varying proportions to convey oodles of information and trigger specific actions. "Virtually all organisms to some extent make use of chemical signals, but the class Insecta raises chemical communication to an art form," observed entomologist May Berenbaum in *Bugs in the System*. And as entomologist Gilbert Waldbauer stressed in his *Insects Through the Seasons*, "Almost all insects depend largely upon their chemical senses, just as we rely mainly on vision and hearing. This is one of the main reasons why humans find it difficult to interpret and understand the behavior of insects."

So much chemical chatter wafts around us in every moment, and we are oblivious to it! We pride ourselves on being a species with superior communication abilities, but it's only in the vocal realm that we excel.

Insects surpass us mightily in the silent and invisible realm of chemical communication.

But back to the luna moth. A year after seeing this ghostly winged splendor at my window, I meet its less glamorous juvenile form. One day while out biking through a forest in Maine, enjoying the pungent balsam and pine scents, I squeeze the hand brakes and lurch to a stop, startled by a neon-green cigar crawling across the black asphalt. Inspecting it closely, I realize that the cigar is actually a giant caterpillar, both long and chubby, each of its dozen chartreuse-colored smooth segments filled out and edged with yellow. Topping the caterpillar are bunches of fine hairs catching the sun and rooted to tiny red knobs jutting from the caterpillar's skin. Two yellow racing stripes run along its sides, but this luna moth caterpillar is no speedster. With its gaudy colors and plump body, it has a comical clownlike appearance, so unlike the ethereal presence of the adult moth. It's like looking back at a picture of the awkward boy at his bar mitzvah, years after he has become a handsome man.

Worried that a car would run it over, and wanting to observe it further, I put the caterpillar in my bike bag and bring it home. By the time I take it out, it has ripened into a russet color and compressed its body like a folded-up accordion. Suspecting it is readying itself to transition into its next life stage (perhaps to escape my prodding), I put it in a large plastic container poked with holes for aeration and lined with leaves it can use to hide its cocoon.

I figure that making a new home in which to overwinter and transform into a moth would be a several-day proposition. But the next morning, looking through the leaves, I'm astonished to see the caterpillar replaced by

• Give a luna moth caterpillar a leaf and it will give you a luna moth cocoon.

a flattened brown blob with a leaf curled around it, presumably tacked in place by the caterpillar's silk. In only one night, the luna moth caterpillar made the cocoon in which it would transform into its pupa, an overwintering sleeping form it takes in between being an insect crawling on the ground to one fluttering in the air. The cocoon is purposely nondescript so no hungry animal will eat it for breakfast.

Who would guess that this drab indeterminate being will one day turn into a stunning numinous moth? One that inspires awe in all of us unsuspecting souls, who never imagine that the world could hold such fleeting fantastic phantoms, battering our windows as if to say, *Let me in to surprise, amaze, and impress you with my brief but beautiful life.*

ASKING WHAT'S IN A NAME

Underwing Moths (*Catocala* spp.)

LOCATION: Windowpane in northeast Maine

DATE: August 20 TIME: 9:43 p.m.

You could write a story based on the actual common names of underwing moth species: The *girlfriend underwing*, despite being a *charming underwing* and not a *gloomy underwing*, soon became the *inconsolable underwing* and *dejected underwing* because she never became the *betrothed underwing*, let alone the *married underwing* or the *mother underwing*. Instead she became the *tearful underwing* and *sad underwing*. She took a lover but then betrayed him by becoming the *Delilah underwing*, with some calling her the *sordid underwing* after that episode. Having gone through a stage as the *penitent underwing*, she eventually foreswore males

• A once-married underwing moth like the one that appeared on my kitchen window. It is part of a family of underwing moths, all of which have amusingly fanciful names.

and became the *Sappho underwing*. By the end of her life, she settled into her final role as the *serene underwing*.

There are 250 species of underwings, with names often more colorful than their appearances. I first became acquainted with underwings when a *once-married underwing* appeared one night on the outside of my kitchen window, staring at me while I washed dishes. A nice evening companion, it impressed me with its three-inch wingspan and striking splotches of scarlet and ebony peeking from underneath drab gray-and-black mottled wings.

"Names help us relate to a species, see it, notice it, care about it."

The moth's appearance explained the *underwing* part of its name, but why *once-married*? Delving into the matter, I discovered that eighteenth-century Swedish biologist Carl Linnaeus, the father of modern taxonomy, gave this large moth its matrimonial name and christened two other underwing species the *married underwing* and the *wife underwing*. Other entomologists followed suit, bestowing these moths with such fanciful names as the *consort underwing* or the *little nymph underwing*.

To name an animal is one step toward knowing it intimately and appreciating how uniquely delightful it is compared to all other beings, like recognizing someone you know in a crowd. "Names help us relate to a species, see it, notice it, care about it," as journalist Brooke Jarvis noted in a *Wired* article.

Naming species is also a way to catalog and protect the rapidly shrinking diversity of life on the planet. You can't protect what you don't know exists. Scientists estimate that only about a thousandth of all species on the planet have been named. I often hear entomologists claim they could probably find a new insect species in their own backyards if they looked hard enough.

All of this imbues the naming of insects with a sense of urgency. But it's also the whimsy of the entomologists doing the naming that impresses me. According to Linnean nomenclature, all species are given a genus name followed by a species name. For example, all underwing moths have the genus name *Catocala*, Greek for "beautiful underneath," and the complete species name for the once-married underwing is *Catocala unijuga*. Amusingly, one entomologist struggling to name a beetle difficult to distinguish from its close kin and with the genus name *Agra* gave it the species name *vation* so it would be called *Agra vation*. There's also the fly *Dicrotendipes thanatogratus*, whose species name translates in English to "grateful dead" and was coined by a Deadhead. There are also the wasps *Polemistus chewbacca* and *Polemistus yoda*, named after two well-known Star Wars characters. One self-proclaimed feminist biologist named a new species of praying mantis *Ilomantis ginsburgae* after US Supreme Court Justice Ruth Bader Ginsburg, who relentlessly fought for gender equality.

Entomologists name new species of insects based on a constellation of consistent traits, many centered on the genitalia, mouthparts, and other minute features often seen only under the microscope. But there has been an eternal debate between those who would group variants into the same species and those who prefer to split them up based on fine distinctions. As Jarvis explained, "The history

• This species of praying mantis, *Ilomantis ginsburgae*, was named after US Supreme Court Justice Ruth Bader Ginsburg.

of taxonomy includes a long series of battles—driven by evidence, opinion, and personal predilection—over whether groups of specimens ought to be lumped together or split apart."

These days many scientists use DNA and not an insect's appearance to decide whether it's a new species or should be lumped in with an already named species. These researchers uncover key sequences of insect DNA that act like the black-and-white barcodes that identify products scanned at a grocery store. They then deploy an algorithm that pronounces an insect a new species if its genetic barcode differs substantially enough from that of related insects. Canadian biologist Paul Hebert pioneered the idea of using genetic bar coding to identify species in 2003, then two years later created a database of genetic barcodes and the species they represent that entomologists use to determine if they have a new kid on the block.

You'd think such technology would have ended debates between lumpers and splitters, but now there is heated debate between barcoders and more traditional taxonomists. Part of that debate stems from the fact that when the algorithm detects something different than what's already in the database, it may not necessarily mark a new species, but rather a member of an already named species that hasn't yet had its key barcodes determined and added to the database. Traditional taxonomists also point out technical errors and blind spots that can make DNA bar coding inaccurate.

But splitters emphasize that results from DNA bar coding suggest the traditional gross appearance-based analyses used to determine insect species aren't precise enough. DNA bar coding, for example, split a well-known skipper butterfly species into ten different species and has also revealed many new species of wasps. One 2016 bar coding study nearly doubled the number of insect species in Canada. The splitters argue that the currently limited sections of DNA analyzed for bar coding may not be sufficient to discover all new species, and once other sections of DNA are also analyzed and entered into the database, even more species will be uncovered.

Some scientists propose combining DNA bar coding with more traditional anatomical measures, especially if that anatomical study is aided by artificial intelligence (AI). Such AI can discern species based on subtle differences in appearance to which humans may be blind. I use that sort of AI every time I ask iNaturalist's pattern recognition system to identify a species. Its algorithm considers how much an animal or plant in my photo resembles any of the millions of photos of animals and plants in its database, as well as what species are likely to be in the area where the photo was taken. Although it continues to improve as AI improves, this software can make mistakes or jump to the wrong conclusions based on the limited amount of information it has. For that reason, iNaturalist relies on others in the amateur and scientific communities to verify species.

• Several different underwing species, each with their own distinctive markings.

Boundaries between species will shift depending on which algorithm or AI program is used and the comprehensiveness of the data on which it's based. And researchers continue to discover new features in insects, such as the pheromones they emit, that may distinguish a species better than traditional anatomical features. As Jarvis noted, "While both morphology [appearance] and genetics can tell us a lot . . . there will always be parts of other organisms' lives that matter very much to them but are hidden from us." She pointed out, for example, that many insects can see ultraviolet light, which we can't, and so they might look quite different to each other than they do to us. As Scott Miller, the curator of butterflies and moths for the National Museum of Natural History, told her, we need to "look closely at these organisms and think of it from the way they think about themselves," rather than try to squeeze them into our own limited frame of reference when deciding whether an insect is a new species.

While reading up on the once-married underwing, I discover a well-known splitter, the nineteenth-century British entomologist Francis Walker. He named the once-married and six other underwing moth species. Walker also gave as many as six different names to what experts fiercely agreed was the same species. One colleague cynically called this splitting-apart practice Walkerism. While cataloging all the insect species in the collection of what is now called the Natural History Museum in London, Walker wrote sixty-seven volumes totaling almost 1,700 pages, in which more than 10,000 of the more than 46,000 species he documented were described as new species. It was his lifetime achievement and took decades to complete, with his poetic naming business perhaps inspired by the poet Lord Byron, with whom his family vacationed. But in his haste to cover so much ground (so many species, so little time), Walker was sloppy and didn't deliberate enough, leading one entomologist to accuse him of "describing the specimen, not the species." *Entomologist's Monthly Magazine*'s obituary of Walker described him as "having done an amount of injury to entomology almost inconceivable in its immensity." Not something you'd want on your tombstone.

But Walker did generate some great nomenclature. So did seventeenth-century Dutch anatomist Jan Swammerdam. Trained as a physician, Swammerdam created specialized instruments to dissect insects under the microscope, proving much to everyone's surprise that the *king* bee in the hive had ovaries and thus was a *queen* bee, and that the worker bees too were female. So much for the patriarchy. Swammerdam also discovered in insect larvae the cellular buds for the wings, antennae, and reproductive organs of the adults. I adore the name he gave these transformational cell clusters—*imaginal discs*, which suggests that a good imagination can launch you from one stage of life into another.

It also takes a good imagination to envision moths as wives or consorts, as charming or dejected. Let's applaud those entomologists who spent decades hunched over their specimens and microscopes to detail some of the tiniest subjects on our planet and yet could imagine something much bigger, could provide the context for a larger story, could see the whole among the parts, while adding a bit of whimsy and poetry to their scientific musings.

We can all learn from them.

CHANCING UPON A LATE-SEASON VISITOR

Bruce Spanworms (*Operophtera bruceata*)
LOCATION: Windowpane in northeast Maine
DATE: November 9 TIME: 7:29 p.m.

It's a freezing November night in Maine. Curled-up remains of maple and birch leaves cover the ground, the bare branches they leave behind revealing the skeletal forms these trees take in winter. Then a beige speck in the blackness appears at the window beside my front door. The porch light imbues a spectral silvery sheen to its pale, fringed semitransparent wings, which span only about an inch and are scalloped with taupe markings. How could this moth still be alive and active on such a cold night? I notice it, after all, as I am bundling up in preparation for going out to gather wood for a fire. Here I am wrapped in layers of flannel and fleece and eye to eye with

• This Bruce spanworm moth appeared on my windowpane one freezing-cold November night.

a naked delicate moth whose human equivalent would be an Arctic Inuit wearing a bikini outdoors in winter.

Is it really possible we share the same world?

When the moth takes off after I return indoors, confirming that it is indeed alive, I cheer him on as some sort of emblem of vim and vigor in the face of death, like that fluttering moth Virginia Woolf wrote about in "The Death of the Moth": "Watching him, it seemed as if a fibre, very thin but pure, of the enormous energy of the world had been thrust into his frail and diminutive body. . . . He was little or nothing but life." I too am impressed with this tiny packet of animation flitting about when most insects lucky enough to still be alive up here in the north lie hidden and dormant. The moths of summer have long vanished, leaving this surprising hanger-on, despite temperatures in the twenties, my breath smoking in the air. What's he doing out there, so exposed on one of the coldest nights yet this fall, and what could possibly be driving him to be on the wing?

"The moths of summer have long vanished, leaving this surprising hanger-on."

Sex, of course.

Bruce spanworm moths time their flying debut until as late as December, by which time most insect-eating birds have already headed south for the winter and most bats hibernate. So this insect deploys a pretty smart evolutionary strategy to avoid his predators. And I know it's a "he" because the females of this species lack wings, opting instead to produce more than a hundred eggs that weigh them down, making flight unfeasible.

Instead of flying, female Bruce spanworm moths hang out at the base of trees and draw mates to them by emitting pheromones, like gals wearing perfume at the bar. After mating, females lodge their pale green eggs in the cracks of bark or other protected sites in trees. Unlike their parents, these eggs survive winter. By early May (in the northern latitudes), caterpillars hatch from them. They change color as they mature, starting off a pale yellow and darkening to green, olive, or brown with lighter stripes on their sides.

You've probably seen these caterpillars before but called them by a different name: inchworms. That's their generic common name because they are about an inch long and inch their way across the ground by bunching up and then stretching out their bodies. They have fleshy leglike structures only on their front and rear ends—hence their funny way of walking. The family of moths that Bruce spanworms belong to are called geometers (from the family name Geometridae), which is Greek for "Earth measurers." I relish the idea of these tiny caterpillars taking a measure of their giant world during their brief lives.

After hatching, Bruce spanworm caterpillars spin strands of silk so the wind can waft them up to the tree canopy, where they spend about six weeks dining on the buds and leaves of trees, often leaving behind lacy leaf skeletons. Although their feeding can temporarily inhibit the growth of host plants in the years when they're especially abundant, Bruce spanworm caterpillars don't appear to cause lasting harm. And these caterpillars provide food for many birds whose songs we enjoy in spring.

Speaking of birds, the Bruce spanworm has another clever strategy to avoid being gobbled up by them—its caterpillar spins silk that it stitches across the opposite edges of a leaf. This causes the leaf to roll up, hiding the caterpillar while it munches. (See "Reading Insect Lore in Leaves," page 129, for more on this.) But some birds, such as chickadees, have learned that there are tasty treats inside these rolled-up leaves, so it's not a fail-safe strategy.

Without realizing it, I had previously encountered a caterpillar of the Bruce spanworm or a similar moth. I was walking through a pine-and-spruce-scented forest and was suddenly startled to come face-to-face with a tiny inchworm dancing in midair. Coming seemingly from nowhere, it

• A Bruce spanworm caterpillar. You can often see these inchworms dangling in the air by the silk they use to lower themselves to the ground from trees.

dangled right in front of my eyes, the wind twisting and turning its lemon-lime thread of a body. I remember thinking, *How can a crawling creature possibly be in the air?* Then I noticed the caterpillar performed its circus act by hanging from a strand of silk suspended from the tree above it.

In mid-June, the caterpillars use their silk like a lifeline to lower themselves to the ground, where they spin cocoons in which they wait out summer, not emerging as adults until November or December. Imagine that—you're on this planet for only about six months and you don't even get to experience the lush warmth of summer. But if that gives you the opportunity to have sex, hey, maybe it's worth it?

Which brings us back to the male moths, not only surviving cold temperatures but flying around the forest in November like athletes of some extreme survival sport. Other moths and butterflies, like owlets and mourning cloaks, can overwinter in northern latitudes, but most lie dormant inside a garage or bark crevice or some other protected shelter. (Naturally produced antifreeze keeps these insects from freezing to death.) Moths need energy to power up the muscles propelling their wings, and that energy is limited when it's cold out and there's no nectar to tap. That's what makes the flying Bruce spanworm moth in late fall so remarkable. How does he do it?

By design.

He's lightweight, has large wings in relation to his body heft, and has a lot of muscle mass, all of which combined lessen the amount of energy required for him to fly to females. The moth is perfectly crafted for this cold-weather feat, one that provides that thread of life spanning to the next generation.

"There's poetry in procreating in close proximity to the dead of winter."

Shortly after mating, male Bruce spanworm moths take their wings out for one last spin and then die—because there is, in the end, a limit to the amount of cold these tiny, furless creatures can withstand. But there's poetry in procreating in close proximity to the dead of winter; there's inspiration in that driving spirit of life in the form of a miniscule moth still out flying on a freezing night.

CAPTURING THE IMPORTANCE OF A SHORT LIFE

Clemens' Clepsis Moths (*Clepsis clemensiana*)

LOCATION: Bedroom in northeast Maine

DATE: July 1 TIME: 10:07 p.m.

It flutters onto the bedclothes one night while I am reading, its straw-colored wings shimmering slightly under the light. Its lack of distinguishing features and drab color don't stop me from wanting to know what kind of moth it is. I am curious about my newfound six-legged friend, but tired enough to put off getting to know it by name until morning. I catch it in my hands, drop it into an empty pill container, and put it in the fridge. With the cold slowing its metabolism, I figure there's enough oxygen in the vial for it to survive until the next day, when I'll identify it and set it free. Even if that's not the case, I rationalize that many moths live less than

• A Clemens' clepsis moth, like this one, taught me the importance of a short life.

a week. So I wouldn't be depriving this one of that much if I snuffed it out before it had a chance to live out its short lifespan, right?

Wrong.

I totally underestimated the importance of a short life to a little moth, its drive to carry on its lineage, its need to have sex. Because after taking the moth out of the fridge the next day, I'm shocked to discover part of one of its antennae broken off and sticking out of the orange-tinged plastic pill container. There's a white scrape in the plastic in which the broken piece of antenna is lodged and pokes through. This is where the moth must have moved its antenna back and forth in an attempt to scratch its way out. *How could that be?* I wonder, but after touching the tiny antenna poking out, its strength, sharpness, and flexibility impress me.

"I totally underestimated the importance of a short life to a little moth, its drive to carry on its lineage, its need to have sex."

The moth flutters when I prod it but lasts only a few more minutes, not giving me time to set it free. Thinking about how it spent its last night frantically trying to pierce its way through the plastic, I feel guilty. An Australian entomologist felt similar remorse when he accidentally injured a cricket he was studying. After watching the insect curl its head down toward its wound and consume its own entrails, Jeffrey Lockwood wrote in an essay for *Orion* magazine:

> My heart sank. . . . I loved these wild animals, not with the sort of conditional affection that we have for organisms that can return our warmth but with a sort of deep empathy without pity. For we ultimately shared a defining reality: the capacity . . .

> for striking out in fear, attacking in anger, and writhing in pain. I believed that I finally understood what Walt Whitman meant in describing animals in *Leaves of Grass*: "So they show their relations to me and I accept them / They bring me tokens of myself, they evince them plainly in their possession."

This Clemens' clepsis moth also brings me tokens of myself and teaches me a lesson: Even when, or perhaps precisely because, your life is short, every moment is precious and worth fighting for. After I first encountered it, I should have released the moth back out into the night, nameless but able to complete its link in an enduring chain of life passed on from one generation to another. Nameless like we all eventually will be, once enough generations pass. I don't remember the names of any of my ancestors beyond my grandparents, but I am grateful for the spark of life they passed on to me and I will pass on to others. They are nameless but not null. Like the moth would have been—if only I had opened the door and let it fly away.

SHADOWING NIGHTTIME GARDEN PARTIES

Sphinx Moths (Family Sphingidae)
LOCATION: Garden in northeast Maine
DATE: August 2 TIME: 8:48 p.m.

After planting fragrant white-flowering tobacco in my Maine garden, I had an impressive new visitor—one who came only at night.

Usually I plant the shorter purple and magenta tobacco varieties, but I missed the heady scent that was bred out of these flowers in exchange for other features. The tall white varieties produce that lovely perfume, and they do so exclusively in the evening, which is odd given that flowers usually are most fragrant during the heat of the day when bees, flies, and other pollinators are buzzing about. The white tobacco flowers also have unusually long spurs formed by their fused petals, with nectar pooling at

• My tall, fragrant white-flowering tobacco attracted a new visitor to my garden.

• A hummingbird clearwing moth drinks from salvia flowers in my garden.

the base. Both traits attracted this new flower visitor, I realized after reading about it.

Here's how I made its acquaintance: One night someone accidentally left on my house's outside light, which spotlights the garden like an artificial moon. Gazing out the window, I noticed something flitting about the flowering tobacco. It was almost as big as a small bat or a hummingbird. But bats in my area eat only insects, so they wouldn't be drawn to flowers, and why would a hummingbird be flying at night?

It had to be an insect, so I went out to capture it with my camera. But it moved too fast, never slowing enough for me to take a photo. Instead of perching on flowers like a butterfly, it hovered beside them, inserting its exceptionally long strawlike tongue (proboscis) down to reach the flowers' nectar.

Strident black-and-white stripes decorated its stout body and large wings, which were almost as wide as my palm when spread out. The insect's size and behavior reminded me of the hummingbird moths I had seen in my garden during the day. It's easy to mistake these moths for their namesakes because they resemble them both in shape and in the way they hover to feed from flowers. They even have a birdlike tail comprised of bristles that help stabilize their flight.

After identifying my mysterious visitor as a type of sphinx moth in the same family (Sphingidae) as the hummingbird moth, it hit me why the white tobacco flowers emit their sweet scent only at night: This sphinx moth is their prime pollinator, so the flowers are most fragrant at night when the moth is most active. Duh. But how did the sphinx moth discover the flowers in my garden, which is surrounded by a dense forest, and how could it see well enough to feed from them? There was no moon to illuminate beyond my house spotlights, and because of the lack of development in our area, our skies are inky black—dark enough to draw stargazers from across the country.

Following that unexpected, enchanting encounter with a sphinx moth, the next spring I purposely planted more of the scented white-flowering tobacco to attract more of its kind. Then one evening in early August, I turned on all our house's outside lights and took my first nighttime stroll in the garden. Wow, what a moth party I found there! As I neared the flowering tobacco and equally fragrant bee balm, several sphinx moths fluttered away, their orange eyes gleaming in the dark, giving them a spectral Halloween-ish quality. Later, when I looked at pictures of these moths online, I saw that they have distinctive round, amber-colored eyes, with what look like large black pupils. They resemble something you'd see on a teddy bear and not on an insect.

As for how they landed in my garden, thanks to their sensitive antennae, sphinx moths are initially lured to flowers by scent. Once they reach flowers, their humidity and carbon dioxide sensors help them predict which ones have nectar. But as sphinx moths home in on a bloom, visual cues become more important, and they unfurl their tongues for feeding only when they can see the flower. Doing so otherwise would be like trying to thread a needle wearing a blindfold. But how can they see so precisely even in starlight?

It turns out their eyes are uniquely constructed to let in more light, like a camera with a larger aperture. Their brains also slow the processing of light, giving more time for their eyes to collect it, like a slow shutter speed on a camera. Together, these capabilities give them exceptional nighttime vision, making them about a thousand times more sensitive to light than daytime-operating insects. Not Superman's X-ray vision, but a superpower nonetheless. They can even distinguish the colors of flowers at night. And just like bees, sphinx moths can remember the location of a nectar-rewarding flower.

Not only do sphinx moths find their food in the dark, but they perform an incredible circus act to eat, hovering in midair while slurping nectar from flowers swaying in the breeze. "Doing all of that in light levels where I wouldn't be able to see the hand in front of my face, that's a pretty remarkable behavior for an animal with a brain much smaller than a pea," Simon Sponberg, an expert in neuromechanics at Georgia Tech, pointed

• Sphinx moths, like this one, have exceptional night vision, enabling them to feed from flowers illuminated only by starlight.

out in *Smithsonian* magazine. To help them with this challenging task, sphinx moths have special sensors on their heads and antennae that monitor their position and rotational motion. These let the moth make rapid course corrections during its aerial acrobatics that include not only hovering but quick zigzagging and flying backward to avoid predators.

• Many moths, such as this Virginia ctenucha moth, are just as striking as their bigger and better-known butterfly cousins.

Clearly I hadn't been invited to the moth party the night I turned on the outdoor lights to find the sphinx moths, and the moths there didn't appreciate my presence, most scattering when I came close. But I appreciated theirs—who knew so much happens in the garden after dark? I often see several moths hopelessly circling the porch light at night, but I didn't expect such frequent frenzied activity in my entire yard. There must have been dozens of different moths flitting about my flowers.

Unlike their better-known butterfly cousins, most moths work the night shift, but many are just as beautiful. Moths at rest may look drab on the outside, but they can flash bright colors when they spread their wings, and some have brightly colored bodies. The Virginia ctenucha moth, for example, has bland brown wings, but if you look closely, you'll see a striking iridescent turquoise body contrasting with its orange head. Moths "may be likened to some people who seem dull at first sight, but on further acquaintance turn out to be unexpectedly interesting or flamboyant or downright fascinating," wrote Dave Winter in the book *Butterfly Gardening: Creating Summer Magic in Your Garden*. "All have interesting stories to tell if we will only take the time to stop, look, listen, and smell the hidden world of moths and their flowers," added entomologist Stephen Buchmann in a US Forestry Service blog post.

"Half of all insects in the world are nocturnal. So if you want to see them, go outside at night."

Moths are also more abundant—there are fourteen times more species of moths than butterflies. One amateur entomologist I know identified more than 1,700 species of moths in his own backyard, near my home in Maine, over a period of thirty years. I don't usually stroll my garden at night, so I had never noticed these moth pollinators before. But then I read in *The Darkness Manifesto*, by zoologist Johan Eklöf, that half of all insects in the world are nocturnal. So if you want to see them, go outside at night.

That evening, one small nondescript gray-and-white mottled moth near me sedately sipped nectar from the purple coneflower on which it perched. I also saw a much larger dart moth at a nearby black-eyed Susan nectar bar. This one had striking yellow, brown, and black triangular patches on its large overlapping wings, which reminded me of an ornamental cloak befitting royalty, or more likely something you would flaunt to attract a mate. Did this moth don it for the special soirée happening in my garden, a ball of sorts for newly emerged moths with sex on their brains? A moth caterpillar's main goal is to consume as much food as it can, but a moth's main goal is to mate and, if female, to lay eggs. Many have only days to weeks to do so before dying.

I spotted a sweetfern geometer moth decorating a window screen, showing off its new beige wings patterned with black dashes, speckles, and encircled white dots, along with several wavy charcoal lines. It was an incredibly intricate and attractive design for an organism that had just spent months as a solid-green inchworm. Nearby, a tortoiseshell-patterned ocher-and-black leafroller moth showed off its own finery, which was, again, more impressive than what it had had as a young squirt hidden inside a leaf. And a painted lichen moth a bit bigger than a firefly stunned me with its long black stripes, outlined in yellowish-orange, and red epaulets befitting a general. These moths, lured by the garden's perfume, were like a bunch of duded-up horny teenagers happy to escape the confines of their childhood forms and strut their new stuff after dark.

But the moths have to be wary of a major nighttime predator: bats. Millions of years of evolution have given some sphinx moths a weapon to fight back—chemicals toxic to bats that suffuse their bodies. When bats are nearby, moths of these species may emit ultrasonic clicks from the tip of their genitals that, like the skull-and-crossbones symbol for poison, say, *Stay away from me!* It's amusing to watch videos of bats using ultrasound to home in on a sphinx moth and then, once they get close enough to hear the moth's warning, suddenly veering away from it.

Decades ago, it was a sphinx moth at a porch party held by insect physiologist Kenneth Roeder on a warm Texas night that led to the discovery that moths can hear. After observing sphinx moths sipping nectar in a nearby garden, one of the guests asked Roeder if the moths had ears. As

Gilbert Waldbauer recounted in his book *Insects Through the Seasons*, "[Roeder] said that he did not know but took a bunch of keys out of his pocket and jingled them near the moths. The moths responded by taking sudden evasive action." This led Roeder to discover two ears near the sphinx moth's mouth.

Many other insects also have ears in all sorts of places on their bodies. Lacewings have an ear on each forewing and dive to the ground when they hear ultrasound. Praying mantises have a single ear between their hind legs that detects this sonic frequency. And katydid ears are on their knee equivalents. The diverse placement of ultrasound-detecting ears in the insect world and their varying structures suggest they evolved independently in multiple insect lineages to protect them from bats, according to Waldbauer.

Sphinx moths' hovering feeding pattern might be another way they avoid bats. They can more quickly evade these predators if they are already in motion, rather than sitting stationary on a flower to feed. Researchers have noted that sphinx moths perform evasive flight acrobatics whenever they sense something looming above them. I remembered this the next time I stalked them in my

• I spotted a beautiful painted lichen moth, like this one, at night. It's not much bigger than a firefly.

garden and stayed low to the ground, because otherwise they quickly flew away with that erratic flight pattern so typical of moths. Their scattershot flight makes it hard for bats to nab them and contrasts with the straight lines most bees and other daytime-flying insects make toward flowers.

Sphinx moths are among the fastest flying insects, clocking in at more than thirty miles per hour, their fast flight enabled in part by their streamlined shape. Both pairs of their narrow wings are coupled together for efficiency, and their great flying ability combined with their sturdy bodies enables some to migrate long distances, from south to north in the spring, and also across the ocean, feeding on and pollinating island flowers that depend on them for procreation. "[Sphinx moths] tend to move pollen farther than bees or birds. That helps plant populations remain viable in the face of habitat degradation," noted Dr. Robert Raguso, a biologist at Cornell University in a blog post by *Smithsonian* magazine.

Flowering tobacco plants rely heavily on sphinx moths to pollinate them. They inspire fidelity from the moths by providing a nectar tube perfectly matched in length to the tongue (proboscis) of the sphinx moth species that pollinates it. That distance varies quite a bit among different flowering tobacco species, and a sphinx moth has a harder time pollinating a flower with a floral tube shorter or longer than its tongue. This Goldilocks situation provides a win-win for both flower and moth—the flower benefits because the moth tends to deliver only pollen from a member of the same species, and the moth doesn't waste precious energy trying to drink from a flower whose nectar it can't easily reach. An added plus for the moth is that it can outcompete less specialized pollinators. That's important when your wings beat more than thirty times a second. This activity requires lots of refueling at nectar stations.

But the downside to such a monogamous marriage between insect and flower is that when one of the partners goes missing, whether due to pesticides, climate change, habitat destruction, or other factors, the other will probably disappear as well.

After reading about sphinx and other moth pollinators, for an hour or so after it started getting dark I would turn on the outdoor lights and prowl the garden. Lurking behind ironweed and other tall flowers, I tried to capture moths with my iPhone camera, doing my best to ignore the mosquitoes biting through my leggings. But most moths are too busy to pose. They hover or flit their way around the veronica while vacuuming up their nectar. They poke their tongues inside bergamot flowers and flowering tobacco without ever touching down.

I had planted the white tobacco plants so that they would fill my evenings with their sweet perfume, but now they also sweeten my nights by drawing sphinx moths to my garden. There they act like busy bees at a time when many pollinating insects sleep. The moth is so named because its caterpillar raises the upper part of its body and tucks in its head, resembling the Great Sphinx of Egypt. And, like that time-worn statue, it's one of the surviving wonders of the ancient world, a natural world shaped by evolution that stretched out the sphinx moth's tongue, put toxins in its body, gave it a remarkable ability to fly, see, hear, and hover, and provided hours of wonderful entertainment for an amateur entomologist crouching behind her perennials, trying *really hard* to be surreptitious in the garden at night.

UNMASKING INSECTS BLENDING IN

Salt-and-Pepper Looper Moths (*Syngrapha rectangula*) and Many Other Insects

LOCATION: Garden in northeast Maine

DATE: July 4 TIME: 6:13 a.m. (*Syngrapha rectangula*)

It's a sparkling summer morning in Maine. Morning dew bejewels the vegetation, and a gentle coastal breeze wafts in pungent scents of evergreens mixed with shellfish and seaweed. While admiring the golden blossoms of my black-eyed Susans in my garden, I notice a piece of black bark marked with white lichen curlicues resting on one of the flower petals.

And then the bark moves.

Looking closer, I realize the bit of tree detritus is in fact a salt-and-pepper looper moth uncannily camouflaged as a piece of lichen-covered bark. This moth is out of place on a bright yellow petal—probably just taking a

• A salt-and-pepper looper moth blends in with the lichen on tree bark. Many insects are quite adept at blending in to their environment to escape detection from predators.

nectar break before returning to a nearby tree, where it usually hangs out all day, well hidden from preying birds.

Later that summer, while hiking in a forest filled with birch trees, I spot a white moth whose open wings are tinged with blue. I try in vain to take its picture, but it keeps eluding me. As soon as I have it in my sight, it suddenly vanishes. Moments later, it reappears, and I follow its flitting flight until, poof, it's gone again. After this magic trick happens a few times, I look more closely at the area where the moth keeps disappearing. That's when I notice it perched on a fallen birch branch, its closed white wings with black markings blending perfectly with the birch bark.

But that mimic is amusingly outdone by another I see in a text message from a friend: *Do you know what this is?* She sends me a photo of an irregularly shaped cream-colored glob tinged with shades of ocher and brown. I look carefully at this rather disgusting picture.

Bird poop? I text back.

She immediately responds, *I don't think so 'cause it just shook its legs or antennae at me!*

It's a moth called the beautiful wood nymph—a fancy name for something that resembled a splat of poop! When it spreads its wings—one flaxen pair edged in brown, the other dabbed with brown splotches outlined in ocher—it lives up to its name. But when it closes its wings, the markings resemble nothing so much as bird droppings. To further that illusion, the moth stretches out its two front legs, which are enlarged by hairs and colored to mimic a thinner line of excrement dripping down from a larger splat. It's a cover so clever, birds fly right by it, not realizing it's a meal waiting to be eaten rather than one already digested!

Looking like bird poop is a common disguise in the insect realm, one used by several species of wasps, moths, and butterflies. Now I think I see bird-poop-mimicking moths everywhere—although often, on closer

• This Schlaeger's fruitworm moth hides in plain sight on a basil leaf by resembling bird poop.

inspection, I find that they are actually just bird droppings and have a good laugh.

Technically, resembling bird droppings is not true camouflage, which describes the ability of animals to blend into their environment, to disappear into the backdrop. A classic example is England's peppered moth. Prior to the industrial revolution, this moth was most commonly white speckled with black to blend in with white lichen on tree bark. But once coal and air pollution killed the white lichen and stained trees with dark soot, the rarer black variety of the moth proliferated. The tables had turned for this moth.

The list of impressive camouflaging or mimicking in the insect world goes on and on because these animals are so good at it! Consider the camouflaged looper caterpillar, which can hide on any flower by chewing off bits of the blossom and attaching them to its back using silk spun from its mouth. Caterpillars of another type of moth can alter their coloring to blend in with whatever twigs they're on; in this species, even siblings may not look alike if one is on a black twig and the other rests on one whitened with lichen. Even more amazing are certain species of planthoppers that turn camouflaging into a group activity. Each insect resembles a single flower, and when they line up on a stem, they mimic a blooming spike of flowers.

We humans don't have insects' incredible ability to blend our bodies into our environment. Instead, with lots of hubris, we *bend* the environment to better suit our bodies. We add water to parched land to grow our food. We remove water from wetlands to grow our food. We warm the cool air in our homes. We cool the warm air in our homes. We kill our large predators. We kill our small pollinators. We make machines to mow down the habitats of others. We build machines to mobilize us in our own habitat, letting us spread like a metastatic cancer across the globe.

Our bending instead of blending has its limits. Ultimately there's bound to be a reckoning, a point at which our inability to adapt to our environment becomes our undoing. Sometimes I think of humans as dragons warming the planet, destroying everything we touch when we exhale, leaving nothing to sustain us. But a scorched globe may mean more room for adaptable insects to take over—and the meek six-legged creatures are likely to inherit Earth.

The tables will turn for us just as they once did for the lowly beetle. About 250 million years ago, there was an apocalypse—maybe an asteroid collision, maybe massive volcanic explosions followed by acid rain, maybe both, or maybe something unimaginably worse. This catastrophic event killed 90 percent of Earth's living beings in what is known as the Great Dying. Beetles and a limited number of other insect species emerged, victorious, from this global calamity. Most likely, these beetles survived because they fed on decaying plants and animals rather than their living versions, so when nearly everything died, they had a feast. Their short reproductive cycle also let them quickly evolve the new traits they needed to adapt to their transformed world—to blend in. Then, sixty-five million years ago, an asteroid collision prompted the next major wipeout of life on Earth, killing off three-quarters of all species, including all the humongous dinosaurs. But the little and often belittled beetles survived yet again.

These little animals thrived while the big ones died in part because their wings let them travel further to find more favorable environments with less exertion than those trudging on the ground. The ability to fly also helped the survival of the remaining dinosaurs, today known as birds. In the modern era, if a nuclear war ends up killing off humans and many other species, once again beetles will likely survive. Experiments with the flour beetle show it can withstand a thousand times the radiation emitted by the atomic bomb dropped on Hiroshima.

The remarkable adaptability of beetles contributes to their status as one of the most diverse group of animals, with more than 350,000 known species—that's seventy times the number of species of mammals. These insects thrive everywhere except in the open ocean and at the north and south poles. One aquatic species has even evolved the ability to survive being swallowed by frogs. Like Jonah in the whale, this death-defying beetle is somehow immune to the animal's deadly digestive juices and able to make a quick run through its gastrointestinal tract and then trigger the frog's sphincter muscle, a rectum equivalent, to open when it reaches the end.

Once, when asked by the archbishop of Canterbury what he had divined about the Creator in his study of His creations, British evolutionary biologist J. B. S. Haldane responded, "An inordinate fondness for beetles." I too now have an inordinate fondness for beetles, as well as for moths masquerading as bark, flowers, or bird poop. Their mighty ability to adapt and blend in far overshadows their minute size such that they are likely to inherit the future world—one we won't be around to see.

COLLECTING FLOWER FLY SUPERFLIES

Family Syrphidae

LOCATION: Garden in northeast Maine

DATE: June through October TIME: Sunrise to sunset

Some people have a bird life list—a personal record of all the bird species they have seen and identified. Me? I have an insect life list. I will stalk any flying or crawling little critter that comes my way, even if it means getting sucked into the mud in the bay while following it, or scratched by wild blackberry bushes growing in a Maine meadow. Once I learned there were more than 2,000 native bee species abuzz in North America, I became eager to make the acquaintance of as many of them as possible. I snapped a picture of every yellow-and-black-striped winged insect I saw and dutifully posted them on iNaturalist to identify them.

Is it a furrow bee, a leafcutter bee, or a mason bee?

According to the experts, the answer was often . . . *a fly*?

• It may look like a bee, but it's actually a fly—a black-shouldered drone flower fly.

At first, I was defiant. *That can't possibly be, c'mon now, can't you see the distinctive stripes?* But flower flies are incredible mimics. My disappointment at not finding a new bee soon turns to delight because when I look closely, I can detect elaborate and beautiful variations in the body decor of these dapper flies. They really are superflies. Now I want to capture shots of as many flower flies as possible for the virtual collection on my camera—another way to grow my insect life list. My appreciation of flower flies also grows once I discover what they do for us.

Flower flies exist on every continent except Antarctica; there are more than 6,000 species worldwide. More than 800 species reside in North America and are quite abundant. One stunning example is the eastern calligrapher, so named because of the scriptlike flourish of curved black lines that mark its yellow-orange back. But you wouldn't notice these details unless you magnified it, as my camera did, because this fly is only about one-quarter of an inch in length.

Flower flies can be tiny, like the calligrapher, or as big as a bumble bee. They are ace mimics, tricking predators and amateur entomologists (and occasionally the professionals) into thinking they are bees or wasps. Many can reproduce the buzzing sound of a bee. Others can imitate both the appearance and behavior of wasps by raising their darkened front legs to mimic wasps' long antennae. They also hold their wings in a more upright V when resting, rather than spreading them outward as a fly would. It's a canny trick, so good that entomology students have mistakenly included flower flies in their bee or wasp collections; a photo of a flower fly once made it onto the cover of a magazine about honey bees!

Look close enough, and you will be able to see through the flower fly's disguise. Flies have two wings instead of four, as well as shorter, often stubby antennae. They don't have the narrow waists of yellow jackets and other wasps. Even more distinct are flies' larger forward-facing eyes, which comprise most of their heads and usually join together if they're males. Bees and wasps have smaller eyes located on the sides of their heads.

Bees are usually furrier than flower flies, but there are exceptions, like the common brown-and-yellow-striped drone flies I sometimes spot dining on the nectar and pollen of Queen Anne's lace growing alongside the road. These hairy flies mimic male honey bees (drones) both in the way they fly and in appearance. A few bee-imitating species go so far as to lay their eggs in the nests of unsuspecting bees. Once the wormlike flower fly larvae hatch, they feed on bee waste—or even on bee larvae.

Like the bees they mimic, flower flies are excellent pollinators, despite bees getting all the positive press about this. Adult flower flies eat nectar or pollen and fall second only to bees and wasps as pollinators. These flies visit nearly three-quarters of all global food crops and nearly the same percentage of wildflowers. There should be a bumper sticker that says *Flower Flies Feed the World!* Flower flies even show potential for replacing bees as managed pollinators—with no stingers to boot.

Unlike honey bees, flower flies don't commute to and from nests to deliver pollen and nectar to those back home, so they spend more time out in the field pollinating, camping out at night on the undersides of leaves and in tall grasses. Experts have estimated that the two most common flower fly species in the United Kingdom transport about the same amount of pollen as the nation's entire honey bee population. Flower fly species that migrate in the fall or spring can carry pollen over long distances, which furthers the spread of isolated and often endangered plant populations, such as those on archipelagos. Flower fly species are also less prone to being endangered themselves compared to bees because they breed more quickly and exploit wider areas with their migrations. (See "Tracking the Great Migration," page 187, to learn more about insect migration.)

• A typical garden will attract many different species of flower flies. Those seen here are: 1. black-legged gossamer 2. white-spotted pond 3. common 4. eastern calligrapher 5. long-tailed aphideater 6. common drone 7. transverse-banded 8. oblique-banded pond 9. margined calligrapher

Flower flies live for only about a month once they reach adulthood. But there is always a new generation waiting to take their place during the growing season. Different species first spread their wings at different times during the summer and fall. I've had a great time noticing the continual parade of new flower flies strutting their stripes in my Maine garden from June through October. As for my flower fly life list, so far I've seen about a dozen, the unique designs of each worthy of a postage stamp. One fall day I saw five different flower fly species on a single aster plant.

Flower flies often lay their eggs on plants harboring aphids, leafhoppers, or other small high-protein snacks for their young'uns to eat once they hatch. This makes them natural pest controllers. Farmers call the larvae of some species aphid slayers because they consume so many of these suckers. Researchers have found that interspersing flowers attractive to flower flies in rows of lettuce plantings or among apple orchards does a great job of controlling many insects that might otherwise damage their yields. And let's not forget the benefits of those species of detritus-consuming flower fly larvae that spend their lives recycling waste.

So it's no surprise that some companies raise and sell flower flies to aid pollination, for pest control in certain fruit and vegetable crops, or to boost decomposition. As British ecologist Karl Wotton once observed, speaking about flower flies in an interview for *Smithsonian* magazine, "It's hard to think of a more beneficial group of insects. They provide great services—for free." Sounds like another bumper sticker: *Flower Flies Are Superflies.* Who cares if I never get to add a mason bee to my life list when there are more flower flies to meet!

HILLTOPS ARE FOR LOVERS

Some species of flower flies engage in fascinating "hilltop mating." This behavior would be akin to men finding the most popular singles bar in town and then spending all their time there trying to pick up women. The flower fly guys congregate on one of the highest points around (a.k.a. a hilltop), which serves as a landmark mating site. There they wait patiently for gal flies to touch down, sometimes aggressively chasing away other males. They don't let females out of their sight until they've mated with them.

"Their [hilltop] territories serve primarily as display arenas where males advertise their availability and suitability as mates," wrote John Alcock and Gary Dodson in an article on the mating systems of "hilltopping" insects, as they call them, for the *American Entomologist*. Alcock and Dodson noted that many insects besides flower flies practice hilltop mating, especially those that are rare in an area. "The fact that so many scarce insect species find mates on hilltops means that entomologists and others interested in biodiversity should promote the

• Flower flies, like these common drone flower flies, find mates on hilltops and other high points in the landscape.

conservation of these places. Perhaps we need a new bumper sticker—*Hilltops Are for Mating (Insects)*—to make the point that insects, and their habitats, deserve our protection as much as the panda and the polar bear."

ENCOUNTERING EXTRAORDINARY ORDINARY FLIES

Green Bottle Flies (*Lucilia sericata*)

LOCATION: Courtyard garden in Philadelphia

DATE: May 6 TIME: 10:12 a.m.

There's a fly in my courtyard garden in Philadelphia. The kind of fly I would normally swat away without even thinking. This time, I take a picture. Peering over the enlarged image, I see that the insect's metallic turquoise and golden-green body is far more beautiful than the flower it perches on. And its black legs and wings with black veins nicely contrast with its shiny, iridescent torso.

• A green bottle fly perches on a black-eyed Susan. These common flies play an important role in decomposing carrion and feces.

This extraordinary-looking fly is so ordinary that it's called the *common* European green bottle fly, and it is one of the most common flies in North America. Why did I never notice how gorgeous it is before this moment?

I'm sure you've seen green bottle flies too. But perhaps you averted your gaze, because they are typically found in some of the most distasteful places. They're the flies that pile onto dog poop and dead rats on city sidewalks and party on dead porcupines up in Maine. These flies, like many fly species, specialize in digesting decaying flesh and animal waste, or at least their young'uns do.

Despite this seemingly distasteful way to make a living, the green bottle has a lyrical and lovely Latin genus name, *Lucilia*, that makes it sound like a fly you'd meet at a square dance—one wearing a colorful, flouncy skirt, of course, and a satin bow in her hair. *Lucilia* means light—an appropriate name for an insect whose iridescent torso glints so beguilingly. In fact, it's that glinting that draws the guy flies to the gal flies. Sunlight reflects off the insect's metallic body as it flaps its wings, creating a blinking beacon. Males can detect fertile females by how often they flap their wings, because fertile ones flap more slowly than males and older flies of either sex. One study found that males are more drawn to an artificial tiny light flashing at the same frequency as a fly's wingbeats than to an immobilized female green bottle. That probably explains why green bottles mate less often on cloudy days, when there is no such sun-induced flashing. No rainy-day sex for them.

Once she's mated, the female searches for cozy spots on fresh carrion to lay her eggs. Flies have been shown to reach a body within minutes of its soul departing from this world. They home in on the dead from as far as ten miles away, using their antennae, which are loaded with detectors that can sense odor-generating chemicals at concentrations much lower than what we can detect with our paltry sense of smell. Many females lay their eggs on the same carcass.

Hours to days later, the eggs hatch into a mass of thousands of writhing grubs, a.k.a. maggots. This nursery of pudgy cone-shaped white or yellowish larvae creates the collective spit needed to digest a dead body. The nursery also maintains the warm temperatures that are key to the larvae maturing quickly. "Living in another organism is a wonderful environment for the little ones to develop in—it is warm, safe, and there is plenty of fresh food within reach," noted entomologist Erica McAlister in her book *The Secret Life of Flies*. Mature maggots can create so much heat that the temperature within their masses can be fifty degrees Fahrenheit warmer than the ambient temperature. Green bottle larvae can also network in carrion in a mysterious way: If a group of them becomes too large such that overcrowding ensues, the larval mass splits into two groups that move to separate areas of a cadaver.

"The insect's metallic turquoise and golden-green body is far more beautiful than the flower it perches on."

Green bottle larvae development is so precisely tied to time and temperature that forensic experts use them to sleuth out murder victims' time of death. They are especially valuable in that regard because they're one of the first insects to colonize a corpse. How far along the flies are in their life cycle on the dead body, as well as information from other colonizing fly species, can help determine the time of death.

Our dear *Lucilia* larvae not only make a living by eating the dead, but they also work in health care. Doctors use them to treat wounds and infections in their patients, relying on the grubs' knack for digesting only dead gangrenous tissue and leaving healthy tissue alone. Wound healing is aided by the antimicrobials the grubs secrete along with compounds promoting tissue regeneration. Even their poop has a nice sterilizing effect.

They've proven especially useful in treating drug-resistant strains of flesh-eating strep infections. Factories even mass-produce sterile teabags full of live green bottle maggots that can be placed on wounds and infections otherwise resistant to healing. The maggots poke their dead flesh–eating heads out of their mesh enclosures to do their business. Patients find the treatment painless.

So, if these flies are known for their role as decomposers of carrion, what drew one to my garden? Adult green bottle flies, especially the pregnant ones, depend on flower nectar and protein-laden pollen, making them important pollinators. These insects also play a valuable role in recycling waste, with many species feasting on feces. So now every time I dodge a dog turd, I think, *Thank goodness for flies!* As nineteenth-century naturalist Jean-Henri Fabre noted in his *Insect Adventures*, "You think they are horrid, dirty insects; but they are not; they are busy making the world a cleaner place for you to live in." Jonathan Balcombe, author of *Super Fly*, commented, "I shudder to think what this planet would look and smell like if we didn't have flies around to clean up the messes other creatures inevitably leave behind when they poop or perish."

So we're lucky the common green bottle is so common. Now I welcome these flying jewels into my garden. "Stay awhile," I tell the one that landed on a petunia petal. "And by the way, while you're here, do you mind taking care of that dead mouse our cat left us?"

• When you look at this green bottle fly up close, you realize how beautiful it is.

DECIPHERING COURTING DANCE FLIES

Families Hybotidae and Empididae

LOCATION: On an oxeye daisy (*Leucanthemum vulgare*) in a garden, northeast Maine

DATE: June 28 **TIME:** 4:17 p.m.

It's a warm sunny morning in Maine, and the bees are busy buzzing about my garden, their humming adding to the thrumming of lobster boats speckling the bay. I take a photo of one of the bees drinking from a daisy. But when I enlarge the photo, I notice beside the bee a much smaller insect—a tiny humpbacked fly with long, spindly legs and yellow-edged feet, sporting a yellow plume on its head like a Roman soldier's helmet. Its long mouthparts dangle down like an elephant's trunk—an extremely tiny elephant. So small is this insect that its golden slippers are *grains of pollen*, as is its striking yellow plume.

And that is how I met my first dance fly.

• A dance fly, no bigger than a gnat, wears golden slippers made of pollen grains.

Dance flies were initially known as balloon flies and first written about by Carl Robert Osten-Sacken, a nineteenth-century Russian baron. While walking through a fir forest in the Swiss Alps, he came across a multitude of sparkling silver specks zigzagging in and out of morning sunbeams. After netting a few and examining one under a magnifying lens, he was "astonished to find a much smaller fly than I had expected, and without anything silvery about it," he wrote in an article published by *Entomologist's Monthly Magazine* in 1877. Next to the fly was a white gauzy ball "so light that the faintest breath of air could lift it." After capturing more flies and scrutinizing them, he found most of the flies clutched these silken balloons under their bodies. He guessed it was these balloons that were catching the light, not the flies, but why were they carrying them?

"How can we understand their behavior when they perceive the world so differently than we do?"

The first to offer a theory was American entomologist Edward Kessel. In the 1950s, after much study of multiple dance fly species, Kessel claimed, in a 1955 issue of *Systematic Zoology*, the males would bring potential mates "a wedding present in the form of a juicy insect," or merely the inedible husk of a formerly juicy insect that the male wrapped in silk extruded through hollow hairs near the swollen base of his forelegs.

Kessel guessed that these nuptial gifts were an attempt to stave off being eaten by the female during mating. Dance flies are predatory carnivores, and it was assumed that the male flies were the hunter/providers. The theory was that females ravenous for the protein they needed to make their eggs more viable might be hungry enough to eat their mates—hence the offering of a snack before getting down to business. Kessel suggested that the male dance fly would sometimes shroud the snack in silk so it took the female longer to unwrap it. This would give the male more time to inseminate her. Kessel speculated that this habit evolved to just giving the silk wrapping to females, without anything inside it. This saved the male insect the trouble of hunting prey for anyone besides himself.

This was the prevailing theory for decades. There's only one problem—it's not necessarily true. (For another case of erroneously overemphasized female sexual cannibalism, see "Catching the *Preying* Mantis," page 29.) According to biologist Rosalind Murray of the University of Toronto, no one has documented such cannibalism in dance flies. There's also no evidence that females are unable to hunt their own protein-laden prey—or indeed if they even require additional protein for their eggs. And the female flies aren't all patsies. Studies have shown that female dance flies end mating once they discover they've been given a sham gift. But often that's long enough for these cheaters to pass on their less-than-desirable lineage. "Even one offspring made by a cheater is some reproductive success for a male that sucks at hunting. He can find a twig and wrap it up and hand it to the female and in that way transfer some sperm, passing on the genes of a cheater to the next generation. So cheating will sneak into a lineage even when detrimental to the progeny," Murray told me.

So why do the female dance flies accept the empty silk packages? One hypothesis is that the male's silk-making ability serves as proxy for some other trait that indicates his offspring will be more successful. Murray noted that it takes time and calories to make silk balloons. "A starving male can't spend time making these gifts," she commented in our conversation. So perhaps the silk balloon is the offering of a successful hunter who can afford to be a fly of leisure. Another hypothesis is that the glinting silk balloons capture a female's attention.

It's tempting to anthropomorphize—to take a sparkling ball of silk and turn it into a glinting engagement ring, or to assume, as Kessel did, that the gender roles of our time reflect a biological pattern. As Hugh Raffles stresses in his *Insectopedia*, we know so little about an insect's point of view. He muses, "What if . . . we assume that those female flies know what they are doing? What if those empty balloons actually are gifts—but gifts whose value we don't understand? . . . Is it too obvious to point out (again) the hazards of presuming what an object is and what it does for beings whose ways of being are so different from our own?"

As much as we may try (see next chapter), we can never fully get inside the mind of an insect or any other animal. How can we understand their behavior when they perceive the world so differently than we do? The next best thing is to avoid projecting our own human-laden expectations on them. We can tease out with experiments and observations what is likely or not likely to be true. But we also have to accept that there are some mysteries out there we may never be able to solve.

• A male dance fly clutching willow seed fluff to give as a nuptial gift to his mate.

INVESTIGATING THE MIND OF AN INSECT

Fruit Flies (*Drosophila melanogaster*)

LOCATION: On bananas in a kitchen, anywhere

TIME: Anytime

A friend once told me she wished she could get into the head of her cat. She wanted to see what a cat's interior life was like, and experience the world through his senses. This reminded me of the Charlie Kaufman film *Being John Malkovitch*, in which people find a magical portal enabling them to literally enter the mind of the famous actor, to hear his inner thoughts and experience the world as he does.

I have no desire to eavesdrop on John Malkovitch, but I'd love to enter the mind of an insect and find out what, if anything, it thinks about. Does it have fond memories of a previous meal, plan its next moves, moon over mates? Maybe we shouldn't expect too much—after all, many insects have

• Close-up of a common fruit fly. Recent research suggests these insects are smarter than we might have suspected.

heads not much bigger than the period at the end of this sentence. But could there be more to insect behavior than mindless, instinctual drives?

Yes!, say many researchers these days, as I discover after exploring the scientific literature. Much of this research has been done on the humble fruit fly because it is the workhorse of experimental biology. For over a century, researchers have been investigating this tiny insect for big insights into early animal development, aging, and everything in between. We may be higher on the evolutionary scale, but we still share nearly two-thirds of our genes with the flies circling our ripe bananas. Our shared biology means that fruit flies have starred in more than 100,000 scientific papers and earned scientists seven Nobel prizes. They are, as evolutionary biologist Martin Brookes dubbed them, the "unsung heroes of science." (Like entomologists, these scientists give fanciful names to fruit fly variants: *Methuselah* for those that live longer than average, *Cheap Date* for those susceptible to the effects of alcohol, and *Ken* and *Barbie* for mutants lacking external genitalia. There's even a *Tin Man* variant that lacks a heart.)

"The fly continues to amaze in how smart it really is."

Scientists have shown that fruit flies can learn and remember by presenting them with an odor along with a shock—a traditional conditioning experiment. The fruit flies learned to steer clear of the odor for long periods of time, even when the shocks stopped. Other studies monitoring the electrical or chemical activity of fruit fly brains reveal that like us, flies pay more attention to novel stimuli and can anticipate a repeating pattern. They also suppress or ignore irrelevant stimuli and don't learn as well when they are distracted—by a puff of air, for example. Intriguingly, the chemical dopamine is the kingpin in both fly and human learning.

"The fly continues to amaze in how smart it really is," said fruit fly researcher Dhruv Grover of the University of California at San Diego.

Gero Miesenböck of the University of Oxford agrees. He found that when the fruit flies he'd trained were presented with two odors, they took longer to decide which was stronger when the odors were of similar pungency compared to when they were more obviously different. This suggests fruit flies, like humans, ponder more difficult decisions before acting, rather than acting automatically. "What our findings show is that fruit flies have a surprising mental capacity that has previously been unrecognized," he observed.

Male fruit flies even deliberate their mating decisions, weighing several factors when mulling over options. In a University of Washington study, fruit flies seemed to use a number of visual, chemical, and behavioral cues to rank available females from ten different strains into a hierarchy of attractiveness. These flies then consistently mated with those higher in the hierarchy when offered a choice between two females. "Creating such a ranking system and then making decisions based on that hierarchy is a hallmark of rational decision-making," stressed Daniel Promislow, the study's lead author.

Additional research suggests that female fruit flies deliberate, too, when choosing mates. In one study, researchers tinted males various colors so female flies could visually identify them, and then they had these females watch the males as they attempted to mate with other females. After that peep show, the females chose to mate only with males that successfully mated the females. Female flies not shown the peep show weren't so selective.

Fruit flies also choose how to react when a predator is nearby. Depending on how active they were prior to the predator arriving, and how close the predator is, a fruit fly will either freeze or fly away. And fruit flies alter their egg-laying behavior in the presence of parasitic wasps—they will lay fewer eggs or deposit them in alcohol-laden, overripe fruit likely to be toxic to wasps. "Far from stereotypic and automatic reactions, defensive behaviors appear to be carefully calculated," wrote E. Axel Gorostiza in a 2018 issue of the scientific journal *Frontiers in Psychology*.

Ants and wasps also deliberate when using tools. Funnel ants use bits of leaf, wood, or mud that they hold in their mouths to sponge up food they can carry back to their nests. But when researchers gave them the option of more absorbent man-made sponges, they used them instead. They even broke the sponges into smaller pieces to enhance their usefulness for the task. If their sponging behavior was directed solely by their innate genetic programming, they wouldn't have been able to change their behavior in order to make use of the new and better sponges the researchers offered them.

So the wheels do seem to be turning inside the heads of many insects, including bees, as we saw in the "Fathoming Brilliant Bees" chapter (page 11). Insects may also have some self-awareness, as suggested by ants that were able to recognize their reflections in a mirror. When Belgian researchers painted blue dots on the front of the heads of ants and then placed them before a looking glass, the ants tried to clean off the dots. But they ignored them if they couldn't see themselves or if the dots were placed on the back of their heads and were hidden from their view.

All these findings puzzle me: How can such a tiny brain do so much? Fruit fly brain cells (neurons) number in the hundreds of thousands, while a bee's brain has about a million neurons and a human brain is comprised of tens of billions of neurons. But bigger doesn't necessarily mean better.

"In bigger brains we often don't find more complexity, just an endless repetition of the same neural circuits over and over. . . . To use a computer analogy, bigger brains might in many cases be bigger hard drives, not necessarily better processors," explained animal behaviorist Lars Chittka.

In *The Mind of a Bee*, Chittka noted that only one widely branched cell in the honey bee links detection of a sugar reward to the part of the brain that processes sights and smells. "Consider, in comparison, that the dopamine reward system of mammals contains tens of thousands of nerve cells all conveying the same message. The insect brain has no room for such profligacy," he wrote.

Chittka's simple computer models of potential brain circuits revealed that just four neurons could enable insects to discriminate between greater and lesser amounts of flowers. When I, a human, am in my garden, a zillion neurons in my brain are turning on and off to perceive and make sense of a great deal of information: I'm seeing many different shapes, sizes, and colors while I'm also hearing bird songs and smelling the fragrant forest. A bee doesn't need to take in all of that information to identify the best place to have breakfast. With fewer than a handful of neurons responding only to perceived brightness and perhaps a few other factors, a bee could conceivably determine where the most flowers are in bloom.

Insect brains are also more efficient in using their limited neurons because these cells make more numerous connections to each other than what is seen in other animal brains. Information is written into these connections (synapses). So sparks flying between a small but well-connected set of brain cells may be all that's needed for an insect to respond to its environment.

"Neuron numbers tell us nothing about intelligence," Chittka asserted. What matters more is how these neurons are linked. So when experience modifies the billion synapses in a bee's brain, there are nearly infinite possibilities for learning and memory. Rather than ask how animals with such small brains can do so many clever things, he offers another question: Why are bigger brains even necessary?

A century earlier, the father of neuroscience, Santiago Ramón y Cajal, expressed similar sentiments as Chittka. In 1915, he and his colleague Domingo Sánchez y Sánchez wrote:

> The excellence of the psychic machine does not increase with zoological hierarchy. . . . When the structure of [fish or amphibian] brains is compared with that of bees or dragonflies, the [former] are excessively plain, coarse and rudimentary. It is as if one were to pretend to hold as equals the merits of a rough grandfather clock with the quality of a fine pocket watch. . . . As always in building her marvelous works, nature distinguishes herself much more in her tiny creations than in the large.

Insect brains can also be more efficient because their neurons are highly specialized, more recent studies reveal. Researchers have uncovered more than 8,000 distinct types of neurons in the fruit fly brain, a vast number compared to the little more than 3,000 neuron cell types seen in the human brain. The electric pathway carrying signals from the fruit fly's eye to its brain alone has more than 150 different types of neurons. (The human retina has fewer than 100.) Some of these insect neurons respond to brightness, color, movement, and edges. This probably explains the fly's amazing knack for evading the fly swatter, and perhaps the bee's remarkable ability to distinguish different flower types. Fruit fly brains, and presumably other insect brains, are also structured such that neurons coding for memorized features of a landmark connect to those that take in compass cues. Those links could enable insects to navigate when seeing familiar scenes.

Chittka believes that bees, like all moving animals, must have some ability to think and make conscious decisions in response to their continually changing environment. Otherwise, bees wouldn't be able to focus their foraging efforts on parts of a garden most in bloom, or time visiting those flowers when nectar is most abundant.

Chittka theorized that if consciousness evolved in humans because it's useful to us, it most likely evolved in other organisms with traits and needs similar to ours. "They share with us the difficulties of moving, probing the environment, remembering, predicting the future, and coping with unforeseen challenges," he and his colleague Catherine Wilson wrote in *American Scientist*. If we apply the same criteria for conscious thinking used for much larger-brained animals, they observed, "then some insects qualify as conscious agents, with no less certainty than dogs or cats."

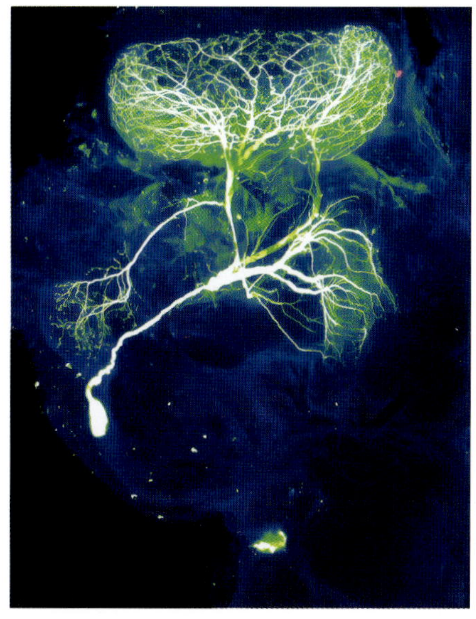

• A single neuron fluoresces in green and white, revealing the numerous connections it makes in an insect brain (blue).

Not all experts are convinced, and the debate will continue until scientists identify the physical substrate responsible for generating consciousness in the brain. It may be that an insect's complex decision-making is subconscious. Further muddying the waters, studies on humans find that we make many of our own decisions subconsciously, without being aware of them. Consciousness can also be a matter of degree, Grover pointed out to me. "I like to think that flies have 'fly consciousness,' that probably differs from human consciousness in that it probably possesses some hallmarks of it, but not all," he said.

FRUIT FLY FEELINGS

For such a seemingly simple creature, the fruit fly does some complex wooing. After tapping a female with his foreleg, the male does a courtship song and dance as elaborate as any minuet. Each species has its own multistep sequence of wing tilts, twitches, or flapping, creating humming songs that would have to be amplified a million times for us to hear. After the song and dance, the male pushes his mouth-equivalent towards the female, licking (kissing?) her before mounting to copulate.

The fly must do this courtship routine in the right order and do a good job at it "or else the entire dance is aborted and the flirting reverts to square one," noted Jonathan Balcombe in *Super Fly*. The fly must also win sparring matches with other males to even have the opportunity to flirt with a female and have a shot at sex. But even males who reach the top of the pack can still be spurned by previously mated and sated females. These gals kick them away or spew out a pheromone that dampens their lust for hours to days.

So it's no wonder that fruit fly males feel dejected and turn to drink when rejected by females. Studies done by Israeli scientists found that unmated males tend to down more alcohol than mated males, suggesting they were drawn to alcohol to improve their mental outlook. Are you skeptical? Well, as Chittka and Wilson asked in their piece in *American Scientist*, "Why would an organism seek out mind-altering substances when there isn't a mind to alter?"

And not surprisingly, male fruit flies seem to really enjoy their orgasms. The Israeli scientists genetically engineered male fruit flies so their exposure to a red light triggers their brain neurons to emit a chemical normally released during ejaculation. The engineered fruit flies spent most of their time in the side of a container lit up by red light. In contrast, normal male and female flies showed no preference for one side or the other.

"I definitely think animals have pleasure," noted Galit Shohat-Ophir, one of the Israeli researchers, in an *Atlantic* article by Ed Yong. "It's hard if you define pleasure from a human point of view, but it [comes down to] very basic machinery that even simpler animals have." It makes sense from an evolutionary standpoint—nature wants all animals to enjoy sex so they continue their lineage.

SPOTTING THE PHANTOM OF THE FOREST

Eastern Phantom Crane Flies (*Bittacomorpha clavipes*)

LOCATION: Windowpane in northeastern Maine

DATE: June 16   TIME: 10:25 a.m.

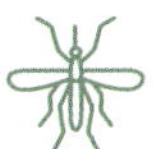

The clip on YouTube is mesmerizing. They bob up and down in the air with their long legs extended so the breeze can carry them like the tufted seeds of dandelions. Playing the now-you-see-me-now-you-don't game, the flies magically disappear and reappear like woodland fairies, their black-and-white coloring camouflaging them in dappled shade.

I haven't been lucky enough to see this vanishing act in person. But one morning in Maine, thick fog dampens my view of the bay and makes me more observant of what it doesn't obscure right in front of me. That's when I spot an eastern phantom crane fly resting on my windowpane. The insect

• An eastern phantom crane fly perches on my windowpane in Maine.

resembles a mosquito but is much bigger—the size of a silver dollar. I admire its lanky beauty. Its legs splay out from its body, like those of a just-born giraffe. Each of its limbs measures about twice as long as its slender torso and bend at the knee and ankle equivalents, like a daddy longlegs. Its striking jet-black body and legs banded with white give it a formal look, as if it's wearing a tuxedo.

I notice spindle-shaped swellings close to the tips of the crane fly's legs. Like the floaties kids wear in swimming pools, these hollow swellings inflate with air to make the fly more buoyant and easily carried on a passing breeze. Crane flies have two translucent wings, but they aren't strong fliers; they rely on the wind for transport. Their black-and-white legs probably help them escape from predators, as they make them hard to see in the forest's dappled sunlight.

The eastern phantom crane fly is one of thousands of species all commonly called crane flies because of their distinctive long legs. You can find them almost anywhere in the world, from the Arctic to the tropics, in deserts, and on the tops of high mountains. I even once saw a crane fly while walking down the sidewalk in the heart of Philadelphia!

Despite their resemblance to their cousin, the mosquito, crane flies are not bloodsuckers, so they don't bite. If they can, they may sip some flower nectar, but most adults don't eat during the days to weeks they are alive. Instead they rely on reserves built up by their more voracious younger selves, chubby grubs that, in the case of the phantom crane fly, nibble on decaying material in the muck of their wetland nurseries. Other species of crane fly larvae dine on vegetation, insects, carrion, or dung.

The sole purpose of adult crane flies is to procreate, and that is usually accomplished within just a day or two of their emergence from their pupal cases. To conserve their limited energy during their fasting or near-fasting adult stage, most crane flies don't flutter about unless pursuing procreation or avoiding predators. So you often see them perched or dangling from leaves and other vegetation or clinging to man-made surfaces like screens and walls.

Because they are so common and frequently immobile, it's easy to spot crane flies. But consider yourself lucky if you can see phantom crane flies floating about in the forest like fairies. One nature blogger noted that she first became acquainted with these insects in Newfoundland when her guide tried to point one out to her. But despite peering intently in line with his finger, she couldn't see it. "I was so frustrated! The bug was flying right in front of me but he was invisible," she wrote. It was only after looking at a photo the guide took of the insect after it had landed that she could see it.

As the blogger wryly commented, "It takes practice to notice a phantom crane fly."

• Despite their resemblance to their mosquito cousins, crane flies are not bloodsuckers, so they don't bite.

READING INSECT LORE IN LEAVES

Moths and Butterflies (Order Lepidoptera), Weevils (Superfamily Curculionoidea), Sawflies (Suborder Symphyta)

LOCATION: Forest and meadow in northeast Maine

DATE: August 1–5 TIME: 9:00 a.m.–4:00 p.m.

We are becoming leaf detectives.

"Look at that line of poop!" my classmate says, holding a bright green aspen leaf up to the sun to reveal a pencil-line-thick black zigzag outlined in white. Who drew this sinuous scribble on an otherwise nondescript leaf, and what can it tell us?

We're taking a field study class on tracks and signs of insects with Charley Eiseman at the Eagle Hill Institute, an organization dedicated to teaching natural history. The institute nestles in a Maine coastal forest filled with aspens and birches along with pines, spruces, and firs.

• A caterpillar has eaten its way through this leaf, leaving behind a thin, black, zigzag line made by its poop.

• On the rolled leaf to the left, you can see the white silk stitches made by the caterpillar inside.

It also is filled with insects, as Charley shows us.

My fellow amateur or professional naturalists have traipsed across the country to take a five-day class focused on these miniscule creatures. We all share a childlike curiosity and wonder about the natural world that most adults seem to have lost. Each encounter with another lifeform—dragonflies, bees, and wasps flying above us; ants, beetles, spiders, and other invertebrates treading below—generates peals of delight and contagious ripples of excitement. "Oh wow!" is a common refrain, along with "That's amazing!" and "So cool!"

We are nature nerds. We stop for wildlife.

Our group can't walk more than a few steps without seeing something intriguing to wonder about, photograph, or try to identify—but it's Charley who helps us notice what our untrained eyes would never spot. "See this curled birch leaf?" he says, his smile spreading in an auburn beard not yet tarnished with gray. "If you look closely, you can see white stitches made from a moth caterpillar's silk. When the silk dries, it shrinks, causing the leaf to curl and give the caterpillar a place to hide while it's eating." He uncurls the leaf, and sure enough, there is a tiny caterpillar inside, munching away, along with a pile of black poop pellets called frass. After we all give our *oohs* and *aahhs*, as if we had seen him pull a rabbit from a hat, Charley shows us a pair of leaves from the same tree stitched together by silk. "This is made by some sort of leaf-tying moth caterpillar," he says. He pulls the leaves apart, and again, there's a caterpillar hidden inside, along with its frass. These critters are cryptic for good reason, as it's not easy being tiny and tasty when you're surrounded by insects, birds, and other creatures eager to eat you.

In that first afternoon of the workshop, we barely enter the woods, so entranced are we in discovering what lies hidden in the leaves of the birches and aspens growing right outside our classroom door.

On our second afternoon, we enter a wildflower-dotted field next to a forest. By the time we arrive, the sun is searingly hot and sweat drips from every pore while cicadas buzz. Even the lichen-painted granite rocks scattered in the field are too hot to sit on. But I forget my discomfort after spotting what looks like a miniature cigar butt dangling from a leaf stem. It's a rolled-up alder leaf, a shelter made by a weevil beetle with a long snout. This critter is less than one-quarter of an inch long, and yet able to create what Charley calls "a masterpiece of insect origami" in which to house her egg.

Later, I watch a time-lapse video of the weevil's process. She begins by biting notches in a leaf's veins to drain it of water and make it more pliable. She then starts rolling the leaf, beginning at its tip, using her legs to neatly fold and tuck in the leaf edges as she rolls. She stops every few rotations to bite wedges in the folds to pleat the leaf and keep the roll from coming undone. It takes about two hours for the weevil to form the spiraled shelter in which she lays a single egg. The rolled-up leaf eventually falls to the ground and becomes both room and board for the hungry grub that hatches. Thanks to the shelter its mother created, this white wormlike critter can devour its leaf in peace, usually unthreatened by nearby predators.

Back in the field, I start eying leaves more closely—what else is hiding inside them?

A lot.

In nearly every tree we look at, as well as in shrubs, grasses, and wildflowers, we spot leaves folded over, curled up on their edges, or bunched together. In them are the larvae of many different moths, butterflies, or sawflies that spin silk strands from their mouths to sew leafy enclosures. If you look closely at the leaves, you can often find the telltale white embroidery tacking down a folded-over flap or rolled edge of a leaf or attaching one leaf to another. The larvae inside usually measure in millimeters.

When I zoom in on my photos of the tiny larvae, I'm surprised by their beautiful coloration: Sawfly larvae sport bright orange heads on lime-green bodies with black stippling; lemon-colored moth caterpillars glisten like wet lollipops, no longer than my pinkie fingernail. Many are so transparent you can see a dark line of excrement inside their small bodies.

"Look how cute this little guy is!" one of my classmates exclaims each time we uncover an invertebrate hiding inside a leaf.

For me, the caterpillars we discover aren't so much cute as confounding. How could such tiny creatures get large leaves to fold over for them or roll up merely by using their silk and legs? I can barely fold a fitted bedsheet by myself, and compactly rolling a sleeping bag has never been my forte—I always struggle to get it back in its sack, despite having hands with nimble opposable thumbs, which insects obviously lack.

• Sawfly larvae munching on the leaf that once hid them.

Charley explains that to fold a leaf in half lengthwise, a moth larva may create several bands of silk across its middle by repetitively whipping its head back and forth. As the silk dries and shrinks, it draws the edges of the leaf closer together. The larva repeats this process, each time laying shorter strands of silk, until the leaf edges touch each other. To create a leaf roll, instead of starting in the middle of the leaf, the larva starts at its edge. It lays down bands of silk that cause the leaf edge to curl when the silk dries and shrinks. It then uses its mouth to sew silk stitches that tack it in place. The larva repeats the banding with shorter stitches until the leaf curls under itself several times.

So a tiny larva speck is capable of rolling an enormous-by-comparison leaf to make its own home. That home may potentially house other critters too. Researchers Camila Vieira and Gustavo Romero discovered that after leaf-rolling larvae leave their verdant homes, other insects and spiders move in, many of them dining on the frass left behind. As LiveScience reported, Vieira and Romero found that "during the dry season, the rolled leaves on 60 plants in the Brazilian forest played host to more than 3,000 invertebrates alone, including spiders, beetles, whiteflies, crickets and many caterpillars."

"Maybe the scribbles on the leaves are a message: *We're here and we'll fill you with awe—all you have to do is notice us.*"

On the third day of our class, we leave for the field early to avoid the midday heat. In a leaf glistening with morning dew, we find a prize: a miniscule gray, brown, and white bird poop–mimicking white admiral butterfly caterpillar. It munches away at a black cherry leaf near the conical leafy remains of its former shelter.

Charley tells us that white admiral and some other caterpillars overwinter in their rolled-up leaves. After eating much of the tip of the leaf, these caterpillars roll up the base of the leaf to form a cozy tube they then line with silk. To ensure that their leaf doesn't fall to the ground in autumn, they also use their silk to strengthen the leaf's attachment to its twig, giving it the appearance of an unassuming dead leaf. I'm surprised. I have often seen the black-and-white admiral butterflies flitting by in summer in my yard, but I never thought to look for traces of them in the winter.

What else have I been missing?

In the field we see evidence of leaf miner larvae, who eat everything but the veins of the leaves and their colorless top and bottom membranes. One of the leaves we later examine under a microscope has a beautiful herringbone pattern that Charlie says are the "teeth marks" of a fly larva. My favorite discovery is an aspen leaf with serpentine patterns, a moth caterpillar's trail of liquid excrement drawing a dark sinuous line as it eats its way through the foliage. A work of art embedded in a leaf.

Charley tells us that during the COVID lockdown, his pandemic project was to discover the leaf miners present in his own yard of just an acre. When he found larvae or pupae of moths, flies, beetles, or sawflies inside mined leaves, he placed the leaves in containers so he could identify the adult insects that eventually emerged. Although many were tiny moths barely noticeable to the naked eye, magnification often showed them to be quite glamorous, with metallic sheens on their bodies or wings topped with furlike collars.

By the end of 2020, Charley had discovered more than 200 species. Two hundred! What abundance lies around us if we just pay attention to it. These industrious tiny critters and the majesties they create are all around us—zillions of leaves sculpted by larvae chewing their way into adulthood, fellow living beings so numerous yet hidden in plain sight. If we could magnify the sound of their mastication, the forest would fill with a resounding symphony. So maybe the scribbles on the leaves are a message: *We're here and we'll fill you with awe—all you have to do is notice us.*

VENTURING ON A NIGHTTIME SAFARI

Fireflies (Family Lampyridae)

LOCATION: Roadside through a forest in northeast Maine

TIME: 8:12 p.m.

My husband and I are on a nighttime safari in Maine, hoping to spot a bear, a racoon, or some other nocturnal creature. Having attended a talk on nature at night, we were inspired to go out and experience it—to explore another realm we didn't know much about. The road we're walking along slices through an evergreen forest sparsely populated with houses and is empty—no cars pass us by. The full moon is as bright as a spotlight; we don't need flashlights, and our long shadows spill out before us. The early fall air is crisp enough to slow the chirping of crickets and deter the mosquitoes that normally keep us from taking such evening walks.

• A firefly beetle rests in my hand.

The stillness sharpens our senses, putting us on high alert. We pursue every rustle in the brush and glimmer in the forest, but we see no wildlife—until I look down at the ground near my feet. There, nestled in the soil, are glowing neon-green embers. Glowworms! These nocturnal animals weren't on our nighttime safari list, making them all the more surprising and reminding me once again that you can behold wonder in some of the tiniest creatures. What are they?

What we call glowworms are actually beetle larvae, the immature forms of firefly beetles, some of whom blink light on warm nights in temperate and tropical regions. Glowworms are found all over the world, although flashing firefly species are mainly found east of the Mississippi River in North America. As a child growing up in a Maryland suburb, I would often head out in the evening to capture these red-and-black bugs, which are close to an inch long. Cupping them in my hand, I would watch them make my hand glow an eerie ruddy color. Then I would put them in a jar by my bed, whose collective firefly bursts of light helped keep the night monsters at bay.

Even as an adult, I have spent many a steamy summer night mesmerized by fireflies' ground-to-treetop display of dancing, blinking lights. I thrill in how their sparks create constellations close to the ground, mimicking those high in the heavens. As Helen Macdonald noted in her book *Vesper Flights*, "We live in a world of distracting, endless glowing screens, but even so these shining, tiny beacons retain an allure that draws people out in droves to stand and wonder."

Japan famously hosts firefly festivals and tours of designated firefly viewing spots, which draw tens of thousands of tourists each summer. At Tatsuno Firefly Park, visitors can walk among as many as a million fireflies performing their flashing light dance, an experience some say is like walking through the star-filled night sky. And in some southern parts of the United States and other tropical parts of the world, you can watch fireflies blinking

• A firefly larva I found clinging to the wall of my house in Maine.

in unison, communicating across the darkness in mysterious communal patterns.

Buried in firefly bellies are "lanterns," which are fueled by a chemical reaction that releases energy in the form of light. The insect's reflector cells, acting like mirrors, magnify the light signal. Entomologist Thomas Eisner called the flashing glows of fireflies "songs in lights." I call them sexual SOSs because fireflies use their flashing lights to find mates in the dark.

Each male firefly has its own species-specific pattern of Morse code–like flashes that it can turn on and off by controlling the amount of oxygen he lets into his abdominal lanterns. (Fireflies invented lighting before Edison and came up with coding before Morse.) The female firefly perches on vegetation and responds with just a single flash followed by a pause. The length of that dramatic pause is also species-specific. When the timing is right, the male firefly finds his way to her. As Eisner put it in his book *For Love of Insects*, "Intimacy in the world of fireflies seemed to be a matter of good timing."

Some species of fireflies contain compounds poisonous to birds and other predators. Eisner and his colleagues found that female fireflies of species that do *not* have these toxic compounds nevertheless find ways to put those toxins to good use, if rather deviously. These nonpoisonous females return the appropriately timed flash that the males of the poisonous species expect. Then, when the guys approach them with salacious expectations, *bam!*, the gals devour them before they even have a chance to realize the ruse.

These unfortunate males have something the "femme fatale" fireflies want: By ingesting them, they also ingest their toxins, which will protect the females and their eggs from predators. "You are what you eat," Eisner noted, pointing out that by eating these poison-laden males, the females take on that species' "mantle of invulnerability." The glowworms we found on our nighttime safari also probably contained distasteful or toxic compounds, with the light they emitted serving as a *Don't eat me!* warning to predators. Even firefly *eggs* glow.

Eisner writes of a frog that didn't get that message and ate three poisonous fireflies. Their toxins quickly killed the frog, but the insects survived for a while in the dead frog's stomach, their flashing light still visible through the frog's skin. "An eerie spectacle," Eisner called it. Eerie indeed. About as eerie as a firefly flashing within the cupped hand of a child, who will use him to chase away all the monsters of her imagination.

• Fireflies flashing in unison create a magical light show in a forest in Great Smoky Mountains National Park in Tennessee in July.

TAKING AN ARMCHAIR SAFARI

Dung Beetles (Family Scarabaeidae), June Bugs (*Phyllophaga* spp.), Japanese Beetles (*Popillia japonica*), Jewel Beetles (*Chrysina* spp.)

LOCATION: Outside of house in northeast Maine (June bug)

DATE: June 11 **TIME:** 9:28 a.m.

An ancient scarab beetle adorns my desk—or at least that's what I imagine when gazing at the June bug carapace sitting there, a more quotidian cousin of the sacred Egyptian scarab. I found this dead beetle at my doorstep in Maine this morning—a June morning, appropriately enough—and brought it inside to study. It measures more than an inch, its long legs bowing out from its smooth chestnut-sheened body.

Is it one of the June bugs that kept flying into my window screen the night before, drawn by my reading lamp? Buzzing and pinging the screen in

• The June bug I found in Maine.

an endless repeat cycle, these insects didn't let up until I turned off the light, their relentless assault making it impossible for me to focus on my book.

Rather than get annoyed by the distracting June bugs, I decide to learn about them. My armchair quest takes me to four continents—Africa, Australia, Asia, and South America—and sends me on side ventures into topics as varied as living jewelry and super-feats of navigation to exceptional parenting and flights of the soul.

The journey begins with the chubby C-shaped white grubs with yellowish-brown heads I had found while digging in the garden earlier in spring. June bugs, which I only tend to notice when they're flying through the sky, actually spend most of their lives underground. It is only when they are three years old that they metamorphose into their adult forms, mate, and lay eggs before dying out by summer's end.

Adult June bugs are creatures of the night, dining in the dark on plant leaves. The morning after they mate, female June bugs bury dozens of eggs in small groupings in the soil. The eggs hatch a few weeks later as root-eating grubs. Some consider these grubs to be pests because in large numbers they can damage lawns, gardens, and agricultural fields. But they're also an important source of high-protein food for other insects, skunks, raccoons, owls, and many other bird species. (Even humans snack on June bugs, in both grub and adult form. Historically, the Indigenous Bear River people of northwestern California ate fire-roasted June bugs, while venturesome folks today cook both grubs and adults, sprinkling these croutons-of-the-sky on salads or crushing and baking them into biscuits.)

June bugs are not true bugs—entomologists use the term *bugs* only for a category of sap-sucking insects—but rather are beetles in the scarab family, which has more than 30,000 species. Like all beetles, June bugs have only one pair of functioning wings because the other pair evolved into hardened wing coverings called elytra. These coverings protect the beetle's underlying hindwings and abdomen when it burrows into soil or explores other sites. But the beetle can still fly with its remaining pair of flight wings if it needs to quickly get away from a predator or find a new food source. Pretty clever adaption, right? It helps explain why beetles are so abundant and ever present.

The June bug's more gaudy cousins include Japanese, jewel, and rhinoceros beetles as well as ancient, sacred scarabs, otherwise and less poetically known as dung beetles.

Few people are fans of Japanese beetles because these invasive insects voraciously devour many garden plants, sometimes to the point of defoliating them. But they do have one redeeming feature—they're gobsmackingly gorgeous. I discovered this by accident one day while out in a meadow in search of insects for my field study class. Noticing something glinting on the leaf of a wild evening primrose, I crept closer to capture it with my phone camera. I was so struck by the beauty of its bright golden-green upper body abutting iridescent copper wing covers that it took me a while to realize this lovely Japanese beetle was the same insect eating my garden plants back at home!

Now, even when I see them munching on my rose of Sharon, I try to think of Japanese beetles as one of nature's jewels, albeit a plant-eating jewel, and console myself with the fact that the damage isn't fatal. They may mar the beauty of some garden flowers, but they also add their own beauty to the petals.

• Japanese beetles like this one can be destructive, but they're also gorgeous when you look at them closely.

Armchair-traveling now to Egypt, I discover that the sacred scarab beetle, a type of dung beetle, was the ancient Egyptians' symbol for their sun god Ra, the supreme life-sustaining god responsible for fertility and crops. Ra moved the sun across the sky each day just as scarab beetles roll perfectly shaped orbs of dung as much as a hundred times their weight with their skinny back legs.

According to Egyptian mythology, at the end of each day, Ra and the fiery ball of the sun sink into the underworld and merge with the god Osiris, who determines which souls to resurrect when Ra and that golden orb return to the heavens at dawn. And wouldn't you know it, after rolling the balls they've made from dung to their burrows, sacred scarab beetles bury them underground for later consumption. (They do this to keep the dung from competitors and to prevent it from drying out.)

In some of these subterranean dung balls, the insects place an egg, which hatches as a grub, which then transforms into an inactive pupa. The pupa resembles an adult scarab but has its legs pressed close against its abdomen, like a mummy. Once it fully matures, the pupa emerges from the earth as a full-grown beetle, like a soul resurrected. Sacred scarab beetles were the first animals to reappear after the annual catastrophic Nile floods in ancient times. "To the Egyptians, the metamorphosis of the mummylike pupa to a new beetle may have symbolized victory over death," wrote Gilbert Waldbauer in his book *Insects Through the Seasons*.

With all that symbolism, it's no wonder that ancient Egyptians, from royalty to peasants, commonly wore amulets representing the beetles around their necks,

• This snazzy scarab beetle is found in southern Texas and northern Mexico.

much like Christians today wear crosses. Hundreds of thousands of these amulets have been retrieved from Egyptian soil, making them almost as numerous as the scarab beetles themselves.

Dung beetles stand out not just as symbols of the soul's resurrection, but as waste managers (see the box on page 154) and as exemplars of parenting in the insect world. One of the most poetic observers of the latter was nineteenth-century naturalist Jean-Henri Fabre, who watched dung beetles both in the wild and in glass structures in his makeshift home lab. He noted in *Fabre's Book of Insects* that the female took great care in burrowing out an underground nest, filling it with a "dung banquet [that] rises from floor to ceiling" and then "modeling its final shape with little taps of her broad feet" before depositing her egg inside. He marveled at its final pear shape, "perfect in outline and exquisitely finished . . . I shall not soon forget my first sight of the mother Beetle's wonderful work. My excitement could have been no greater had I, in digging among the relics of ancient Egypt, found the sacred insect carved in emerald."

Fabre noted that the dung beetle takes care to place her egg inside the narrow, less dense top part of the dung pear she creates, which becomes "a nice airy room with thin walls for her little grub to live in, during its first moments." There the grub is "surrounded by the food that suits it best. It can begin eating at once, without further trouble." More recently, scientists have observed some dung beetle mothers going even further, sticking around for a few months after the grubs have hatched, dutifully cleaning them and taking out their feces.

Other types of dung beetles dispense with rolling a ball of dung to a new spot to bury it underground. Instead, they tunnel through a pile of dung and create their family homes beneath it. (As a real estate agent would say, it's all about location.) The females of certain species drag dung into the underground nests they've created, chewing up and processing the poop they pack into dung sausages for their young. For every egg laid, its mother will make more than a hundred different trips into the ground. In other tunneling species, parental responsibilities may be shared, with both sexes cooperating to create the tunnels and dung loaves that fill them. One or both parents stay with the grubs until they mature, which can take up to four months. Such parental care is unusual in the insect world.

Also impressive are the great lengths to which some male tunneling scarab beetles literally go to ensure their own progeny. After the female excavates a burrow in which to store food for her future family, the male will guard the burrow's opening from being entered by rival guys who could mate with his gal. Males of such species are well suited for guard duty. They have large, teethlike spines on their legs, the better to brace themselves against the walls of the tunnel, and enormous horns that work like the bars of a jail cell to block other males from entering. The horns are huge—sometimes as long as the entire length of the beetle and as much as one-fifth its body weight. "They are analogous to an extra leg on your head [you have to carry] around for your entire life," biologist Douglas Emlen of the University of Montana told Terry Gross in a May 4, 2009, episode of *Fresh Air*.

The largest males or those with the longest horns are more likely to win mating battles. To devote the resources necessary to support such big-horned weapons, other parts of the beetle's anatomy shrink, including their wings, eyes, and, surprisingly, testes.

As Gross noted, "There's this whole world of sex and violence that's kind of invisible to us . . . happening [right] underneath cow manure." In fact, "this whole story is unfolding in most people's backyard[s]," Emlen agreed, as these dung beetles and the poop they thrive in are so common.

Probably one of the best-known examples of horned scarabs are rhinoceros beetles, which resemble their namesake and can be seven inches long. They're among the strongest animals for their size, able to lift up to 850 times their own weight—the equivalent of a person being able to lift a car! Rhino beetles, which include the Hercules and Atlas beetle species, are popular pets in Asia, where you can buy their grubs in vending machines and their food in grocery stores. You can also gamble on which males will win battles when put together in an arena—the insect version of cockfighting.

But my favorite story of these horned species of tunneling scarab beetles is about how smaller males of their kind sometimes still get the girl. How do they do it? Sneakily. After one of these smaller males gets kicked out of a guarded tunnel, at the bottom of which the female beetle is busy compacting and shaping the dung sausages in her burrow for the next generation, he digs his own tunnel right next to it. He then cleverly creates a passageway between the two tunnels below the horned and horny male diligently keeping an eye out for aboveground intruders. Then the little guy scoots in and mates with the female scarab beetle before the brawnier beetle gets a chance to mate himself. Brilliant.

• Above is an Atlas beetle from Malaysia; below is a Hercules beetle from the Southern Appalachians. Their long horns help them guard their burrows.

A sneaky male who wants to get the doo-doo without doing the work can also rob another beetle rolling a dung ball to his home burrow or to give as a wedding gift to a female once he finds a suitable place to bury it. This is tragic when you consider all the work the more industrious beetle does. He uses his flat head with teeth to rake up the dung, toothed legs to scrape and shape, and flattened "elbows" on its bent front legs to press the dung into shape. Once it forms a large, irregularly shaped mass, the beetle gives it a final spin to create a perfect sphere.

To roll the dung ball, the beetle clasps it with its long hind legs while walking backward with its forelegs, keeping its head down and butt up. Home can be fifty feet or more away, and along the route may lie a number of obstacles. An incline, for example, may allow the beetle to get only so far before gravity forces the ball to tumble back down, dragging the beetle with it. "The merest trifle ruins everything; a grass-root may trip him up or a smooth bit of gravel make him slip, and down come ball and Beetle, all mixed up together," Fabre explained.

The beetle engaged in this task seems less like the supreme Egyptian sun god and more like the pitiable Sisyphus. In punishment for cheating death, Hades, god of the underworld in Greek mythology, forced Sisyphus to forever push a huge boulder up a hill only to have it tumble back down when it reached the top, leading him to repeat this onerous task for eternity. But rather than viewing the beetle as cursed, or laughing at its bumbles immortalized online, perhaps we should applaud its persistence, along with its amazing feats of navigation that keep it on the right path. Scientists studying dung beetles in their lab saw both at play when they turned the surface on which a beetle was rolling its dung ball ninety degrees. "This animal actually had the whole world turned under its feet. . . . But it doesn't flinch. It knows exactly where it wants to go, and it heads off in that particular direction," recounted South African entomologist Marcus Byrne in an August 2012 TEDx Talk.

Byrne's study and other research has found that to orient itself and stay on the shortest path back to its home, instead of using landmarks, the beetle uses as its compass the sun during the day and the Milky Way at night. The beetle can also measure how far it's traveled in one direction, which, combined with the celestial landmarks, helps it navigate. (Yet another animal that has a better sense of direction than I do.)

Unfortunately, after a beetle has toiled to make a perfect sphere of dung and expended a great deal of energy and patience rolling it a long way, the poor industrious guy often gets hijacked by a lazier male and loses his dung ball to this ne'er-do-well. Some of these thieves are sly, seemingly helping a fellow dung beetle push his sphere of food along, but doing little of the work, essentially following along, biding his time until the scarab beetle begins digging a burrow for his own dung dining room or for his gal to lay an egg in. Once this burrow gets deep enough that the hardworking beetle disappears into it, the beetle thief "seizes his chance, and hurriedly makes off with the ball, which he pushes behind him with the speed of a pickpocket afraid of being caught," Fabre reported. "If the thief has managed to get safely away . . . the owner can only resign himself to his loss, which he does with admirable fortitude. He rubs his cheeks, sniffs the air, flies off, and begins his work all over again. I admire and envy his character."

My journey into the world of the far-ranging and talented family of scarab beetles, from jewel scarab beetles to June bugs and Japanese beetles, has taught me the value of persistence, how brains can win over brawn, and how beetles can be both annoying and beautiful, destructive and creative, earthly and godly, symbols of both the divine and the cursed. Humble heroes that gotta do what they gotta do to make a living and resurrect the next generation.

BEETLE BAUBLES

Exquisite jewel scarab beetles, resembling June bugs dipped in gold, glow like drops of sun on the forest floor in the southwestern United States and Central and South America. I once came across a display of several beautiful pendants made from the wings (elytra) of these jewel scarabs and was tempted to buy one, but I decided I'd rather have a living jewel remain in its natural homeland rather than a dead one roped around my neck. Then I discovered *living* brooches made from ironclad beetles. (Remember the indestructible diabolical ironclad beetle in "Catching the *Preying* Mantis," page 29?) After being adorned with rhinestones and pearls and attached to a golden chain and pin, these bedazzled beetles wander about on the clothing they're pinned to. Because ironclad beetles naturally reside in the desert, they'll survive for up to three years kept in a jewelry box with just a bit of fungus to eat.

My friend Adele almost had accidental living jewelry while working at the Cincinnati Zoo's insectarium. After strikingly patterned African scarab fruit beetles on display there seemingly died, she took them to a jeweler to make them into earrings. But after she put the beetles in the jeweler's

• A Victorian-era double scarab beetle gold brooch.

hand, they suddenly started moving, apparently stirred from their hibernation! So much for beetles adorning her ears. Adele returned them to the zoo.

IMPORTANT WASTE MANAGERS

The antennae of dung beetles are exceptionally good at detecting animal waste from dozens of miles away, and their fast flying lets them arrive quickly to digest it before it dries out, becoming unpalatable. Douglas Emlen described to *Fresh Air* host Terry Gross collecting elephant dung he found on a road in the Serengeti and bringing it back to the edge of his camp to see what dung beetles it would attract:

> Pretty quickly it was impossible to see anything because they were all buzzing and circling around our headlights and then a couple of minutes later they were starting to pour out of the sky so fast that I couldn't even take notes on my clipboard because my clipboard was covered with literally an inch of solid beetles. . . . It was as if somebody stood over us with a bucket full of beetles and poured it steadily out on top. The beetles absolutely covered

the elephant dung: Tens, probably hundreds of thousands of them came within minutes.

• A dung beetle from South Africa rolls his ball of dung with his rear legs.

Australians know well the important role these turd tumblers play in recycling animal waste. After the English introduced cattle to Australia, their manure began piling up in the fields along with blood-sucking insect pests, choking off the plants on which the cows fed. This manure took months to years to decompose. That's because, although Australia has its own species of dung beetles, they don't digest cow manure: These native dung beetles prefer to dine on the small, dried waste pellets of kangaroos.

This caused a costly and smelly problem Down Under—until the Australian government imported cattle dung–eating scarab beetles from other countries. These insects quickly adapted to their new environs (and old food source) and began recycling the manure back into the soil within two days of it being deposited on the ground.

One US study has found that dung beetles, by boosting the amount of nitrogen in the soil, increasing grazing sites, and reducing dung-breeding parasites of cattle, could add about $2 million to the economy each year.

PROBING LIVING DEAD TREES

Flower Longhorn Beetles (Subfamily Lepturinae)

LOCATION: On an oxeye daisy in a garden, northeast Maine

DATE: July 7 TIME: 10:53 a.m.

It's depressing to look at.

One spring we return to our Maine abode and discover that at some point over the winter, a doozy of a windstorm had toppled a number of trees, making our forest look like a game of pick-up sticks. I worry about how long it will take the woods to recover from the storm. Those towering trees used to pull our gaze upward, but now so many are on the ground or hung up in their neighbors, distracting and disrupting our previous uplifting view.

A few days later, I spot an insect that, I soon discover, might come to the rescue. This beetle decorates the yellow center of a wild daisy that seeded itself in between asters and black-eyed Susans in my garden. She is

• A flower longhorn beetle (*Evodinus monticola*) decorating a wild daisy in my garden.

about half an inch long and svelte, with a broad-shouldered torso tapering toward the back. Her long legs, extremely long antennae, and small head are all in black, as are the striking spots on her golden body, which softly shines in the sun.

> **"Flower longhorn beetles do a great job of cleaning up deadwood, restoring their nutrients to the soil to feed future trees."**

I'm thrilled because beetles can be hard to spot. Many are tiny and inconspicuous, or they are secretive, hiding under logs or burrowing into the soil or trees. But flower longhorn beetles, like this *Evodinus monticola* one I have found, feed on flower nectar and pollen, explaining why I see it on the daisy. Once she finds her perfect tree nursery (see the box on page 160), this flower longhorn beetle will lay her eggs under its bark. The eggs will hatch into wormlike larvae that munch on the woody fibers surrounding them, with the fungi and bacteria there aiding their digestion.

E. monticola grubs prefer to dine on already-dead trees and, along with other insects, worms, millipedes, and fungi, do a great job of cleaning up deadwood, restoring their nutrients to the soil to feed future trees. Such deadwood can house thousands of wood-boring beetle larvae, noted Norwegian entomologist Anne Sverdrup-Thygeson in her book *Extraordinary Insects*. "There's plenty of sugary sap beneath the bark, and when it ferments there's a real party atmosphere among the visitors," she wrote.

I had fun imagining drunken beetles weaving and bumping into each other. Each type of wood, from bark to sapwood to heartwood, has its beetle specialists who greedily gorge on it, and "Once fungi and insects, mosses and lichens, and bacteria have moved in, there are more living cells in the dead tree than there were when it was alive. So ironically enough, dead trees are actually among the most living things you can find in the forest," Sverdrup-Thygeson wrote. She stressed that deadwood promotes biodiversity and called for protecting it rather than quickly clearing it away, as often happens in a managed forest. Many countries have launched initiatives to protect forest deadwood, which experts estimate should comprise almost a third of woodlands.

Some of the flower longhorn beetle's cousins, such as the Asian longhorn beetle, damage trees and other types of plants. But most of these destructive longhorn beetles aren't native to the areas in which they cause damage, so they lack the predators and diseases that keep them in check like they did back in the old country. The vast majority of longhorn beetles feed only on already damaged or dead trees.

As I digest all this information about longhorn beetles, I think about all the deadwood needing to be digested in our surrounding forest. Our neighbor always encourages us to clean up our forest, clearing it of such debris. But out of a combination of ecological-minded altruism and laziness, we let nature take its course. Deadwood provides habitat not only for insects but for other animals I like to see and hear, including salamanders, snakes, squirrels, foxes, raccoons, woodpeckers, owls, and hawks.

So, despite the chaotic look of the fallen trees the winter's storm left us, we only minimally clear them, sawing up and carting away just those that impede passage on our long gravel driveway. Seeing so many trees blown down makes me sad, but now I imagine all the bustling life inside of them, and I think about how the storm has been a gift for so many animals. I hope I'll see more flower longhorn beetles decorating my garden, signaling that more grubs are digesting the deadwood in our forest. There's a lot that needs to be cleaned up out there, so please get to work, guys!

FINDING THE PERFECT BEETLE HOME

The key to the longhorn beetle finding a nursery for her young is her long antennae. This beetle is mighty picky about where she lays her eggs and will opt for only certain kinds of dead trees with certain kinds of damage. Trees release specific chemical compounds when they are stressed or damaged, and the beetle's antennae are chockful of various sensors that help her detect these chemical cues, revealing both the identity and quality of potential host trees. Yet another insect superpower!

Some longhorn beetles' antennae even have smoke detectors, which, in combination with their strong flying abilities, allows them to zoom in on trees recently killed by fire, even if they are many miles away. The antennae can also detect aggregation pheromones exuded by bark beetles, which are usually the first ones to colonize dying or dead trees. The pheromones draw other bark beetles to a good food find; but one species of longhorn beetle cleverly uses her detection of these pheromones to lead her to a good site to lay her eggs, whose hatching larvae have

• This longhorn beetle's long antennae are chock-full of sensors to help her find a good home for her young.

been noted to sometimes ungratefully prey on the bark beetles that drew their mother there.

As Sverdrup-Thygeson pointed out, "When insect mom is house hunting in the forest, her priorities are different from ours. . . . Whereas we fear moisture damage and rot, beetles [that live on dead trees] think they're fantastic because they're like a fridge full of food for the family's greedy kids."

UNEARTHING SUPERORGANISMS

Leafcutter Ants (*Atta cephalotes*)
LOCATION: Forest in the Peruvian Amazon
DATE: April 15 TIME: 7:22 a.m.

When I was a young child and my center of gravity was closer to the ground, I loved watching ants in action. Placing obstacles in their way, I tried (unsuccessfully) to disrupt their relentless conga lines toward food sources. I watched with big eyes and a small person's envy as they carried large burdens—food crumbs and insect prey much bigger than themselves—and I was touched to see them cart away their dead brethren.

On rainy days when I couldn't observe ants, I spent hours drawing detailed imaginings of them living in their underground colonies—ants cooking in kitchens, ants watching TV in living rooms, ants sleeping in bedrooms, even ants corralled in cribs. Eventually, I outgrew my fascination

• Leafcutter ants carrying the leaf shards they use to cultivate fungus back in their nest. The smaller ant riding the leaf protects the larger ant carrying it below from deadly flies.

with the ant world and began to see these insects as annoyances underfoot rather than marvelous beings.

That changed decades later when I visited the Peruvian Amazon.

In a steaming jungle, after craning my sweaty neck to get a glimpse of feathered jewels flitting in dense vegetation above me, I happened to look down. I was startled to see a train of leafcutter ants winding around my feet, parading chartreuse shards of foliage above the rust-colored soil. I had always been charmed by these ants when I saw them on nature shows, but seeing them in real life was like seeing a movie star in the flesh walking down the street. "Oh my God, look!" I shouted to my husband.

These leafcutter ants don't have the kitchens and TV rooms of my childhood drawings, but they are as industrious as any human—they are farmers, each with their own specialized task in the farming operation. The leafcutters that are foragers cut and carry leaves back to the nest, where gardener leafcutters chew them into a mulch fertilized with their own feces. This mulch supports a fungus that is totally dependent on the ants' tending and grows nowhere else in nature outside of leafcutter ant colonies.

Leafcutter farming operations include other types of workers: those that harvest and distribute the fungus to colony members; the supersoldiers, whose bulging heads and strong, sharp jaws viciously defend the farm and nearby terrain from intruders; and, perhaps the most incredible of all, tiny bodyguards for the foraging ants. As the foragers make the long walk back to the nest, they are besieged by parasitic gnats. The ants are unable to defend themselves as they carry their fungal fodder, and the gnats lay eggs on their necks—eggs that hatch into grubs that burrow into and eat the ants' brains. So, to defend foraging ants from these gnat swarms, smaller ants from the colony accompany the larger ones on their foraging trips. On the walk home, these tiny ants ride the leaf shards, using their hind legs to kick away any gnats that come close.

Humankind's ability to farm and provide a steady source of grains and other crops enabled our population to multiply exponentially. But for millions of years before we figured that out, leafcutter ants were tending to their fungal crops, making them one of the most successful ants in the tropics. They're so numerous in the Amazon that we could see ant highways—tracks in the forest floor worn down by their teeny-tiny feet. The mother queen of a colony can give birth to as many as 200 million worker ants—more than half the size of the US human population. A single leafcutter ant colony can strip a tree of its leaves or destroy a family garden overnight. Early Portuguese settlers in the Brazilian Amazon claimed that either Brazil would conquer the ants, or the ants would conquer Brazil.

The leafcutter ants' own settlements are awe-inspiring—huge underground cities extending as far as fifteen feet belowground. They are a complex labyrinth comprised of thousands of chambers interconnected by intricate tunneling. The ants constantly carry nesting matter from the bottom of their nests up to the surface and bring down fresh replacements. This turnover dries out the more humid interior of the nest, prevents mold from growing, and provides air conditioning to the underground city.

Leafcutters aren't the only ant species with sophisticated cooperative behaviors. Thatching ants of the Pacific Northwest herd aphids, moving them to new grazing grounds as needed in order to gather up and feed their young a fresh batch of the nutrient-rich honeydew the aphids excrete while sucking on plant stems. When water levels rise around the nests of fire ants in the southern United States, they join their bodies together to create living rafts that float until water levels subside and they can find drier ground. When the army ants of Central and South America encounter a stream or

• A weaver ant uses silk produced by a larva to stitch together the leaves in its nest.

a deep crevice, they form living bridges by linking legs and jaws. And then there are the marvelous tropical weaver ants, who stitch their nests in the tree canopy by shuttling their silk-producing larvae back and forth from leaf to leaf. "Such tool use may be explained away as a series of simple behaviors strung together to create the appearance of careful planning and rational thought, but it still gives one pause with respect to our place in the universe," noted entomologist May Berenbaum in her book *Bugs in the System*.

How can ants' pinhead-sized brains foster such complex behavior? Volume-wise, an ant's brain is a million times smaller than a human brain. But an ant colony of more than a million individuals could collectively have more brain cells, and thus more computing power, than the human brain, noted ant researcher Ewa J. Godzińska. In other words, each individual ant

• A field ant protects these aphids from predators in exchange for feeding on their waste.

has a tiny brain. But these brains linked together can function like a nervous system to accomplish great feats.

Deborah Gordon, a biologist at Stanford University, observed that ant behavior, like a computer's processing, is carried out without any central control. Although ant colonies have queens that lay all the eggs, these queens do not rule over anything beyond the reproductive success of their colonies. "There's no management. No ant directs the behavior of any other ant," she explained in a February 2003 TED Talk.

Gordon embarked on a quest to figure out how ants accomplish such astonishing feats of cooperative and seemingly intelligent behavior without a boss telling them what to do. Since 1988, she has followed harvester ants residing in about 300 colonies in the Arizona desert. Using fiber-optic video apparatuses poked into the ants' nests, she's developed a better understanding of how these ants carry out their complex operations, which include gathering and storing seeds, tending their young, and maintaining their nest.

Gordon discovered that at any given time, about a third of the ants in a harvester ant colony are loafing around and can be put to work doing whatever tasks need to be done. When she scattered toothpicks around the nest entrances of colonies, more ants rallied to the nest maintenance task, pushing the toothpicks to the outer edge of the nest mound. If she put a pile of tasty millet seeds in front of the nest, more foragers went to work. How did the ants assess who was doing what in order to rally to the cause? Did they count how many of them were doing each task?

Sort of.

I watched a video of returning foraging ants scurrying down a narrow dusty tunnel leading from the outdoors into their nest and being greeted by their nestmates, who high-fived them with their antennae. Gordon discovered that the aroma (a chemical signature) that accrues on foraging ants differs from that of those maintaining the nest. Ants use their antennae to detect those distinctive odors, tapping each ant they encounter with their smelling appendages, which explained the high-fiving I saw in the video. "An ant uses the pattern of its antennal contacts, the rate at which it meets ants of other tasks, in deciding what to do," Gordon said. "The pattern itself is the message."

She noted that the human brain works the same way. When you look in your pantry and see it empty, a buildup of electrical signaling from multiple neurons relaying information on what's lacking in the pantry and fridge triggers you to go grocery shopping. Similarly, an ant may decide to go foraging based on how many contacts it makes with foraging ants returning to the nest. No one neuron triggers the behavior, just like no one ant decides how many foragers to send out of the colony. Control is collectively distributed in a communal brain in the colony rather than centralized in a select cadre of ants.

"Using only simple interactions, ant colonies have been performing amazing feats for more than 130 million years," Gordon said in an interview for NPR's *TED Radio Hour* on April 24, 2015. "We have a lot to learn from them."

More than 14,000 known species of ants live in colonies, but there isn't a leader among them. The success of these colonies, which all prize cooperation over selfishness, is a slap in the face of the Darwinian rules of survival of the fittest and most competitive. Instead, team spirit rules.

But there's a dark side to this team spirit: tribalism. Although ants are extremely cooperative and make supreme sacrifices for nestmates *in* their colony, they can be really nasty and competitive to ants *outside* their colony. "Ants are the most warlike of all animals, with colony pitted most violently against colony of the same species," Harvard University biologist E. O. Wilson wrote in *Tales from the Ant World*. "Extermination is the goal for most. . . . Their clashes dwarf Waterloo and Gettysburg."

When Wilson blew the scent of another organism into a nest hole of an ant colony, within minutes, supersoldier ants came running out, ready for battle. They will defend their nests and the territory surrounding them from any organism they deem Other—one that doesn't have the aroma of their own colony, the scent of kin. Sound familiar?

Wilson proposed that evolution and natural selection operate at both the colony and the individual levels in these social insects. Evolution at the colony level can explain much of the leafcutter ants' cooperative and altruistic behavior: their dangerous foraging work to support their nestmates, their willingness to eat less food so others have more, and their suicidal defenses during skirmishes with invading species. It can also explain leafcutter ants' viciousness toward others outside their colony.

All these behaviors enable queens in their colonies to live longer, producing more surviving offspring. "In order to survive, a colony . . . must consist of cooperating elements fitted tightly together in the same way that organs are fitted together to function as an organism," Wilson emphasized. He called leafcutter ant colonies "superorganisms."

Some view Earth as a superorganism. But the tribal nature of the human species continues to impede progress in combating climate change that would protect the health of our planet—and our species. As journalist Farhad Manjoo noted in a *New York Times* column on August 4, 2021, "Deforestation in the Amazon rainforest could well affect the sea level in Florida, but it's probably difficult to forge much common cause between poor farmers in Brazil and retirees in Boca Raton." As he asked, "Are we capable as a species of coordinating our actions at a scale necessary to address the most dire problems we face?"

Yes, answered British scientist and futurist James Lovelock, but perhaps only by creating non-individualistic-thinking cyborgs that wrest control of the world from humans. Such an artificial intelligence takeover with distributed collective coordination may be our only redemption, Lovelock argued in his book *Novacene: The Coming Age of Hyperintelligence*. He claimed that the intellectual prowess and shared consciousness of cyborgs would enable them to rationally recognize that they are vulnerable to climate change. This would prompt them to engineer a solution that could supersede the fragile alliances and competing needs of billions of humans.

In other words, the whole world needs to behave more like an ant colony.

• Two harvester ants smelling each other with their antennae. The ants use the rate at which they encounter other ants to determine their next move.

Not that the ants themselves care. Because of their outstanding social structures, ants will probably fare better than most if an irreversible and calamitous climate change ensues, Manjoo noted. "Which, really, is no surprise," he said in another *New York Times* opinion piece, this one from November 18, 2022. "Ants were here before us, and they are likely to long outlast us. They run the place. We're just visiting."

PURSUING SISTER POWER

Black Carpenter Ants (*Camponotus* spp.)
LOCATION: Great room floor of home in northeast Maine
DATE: August 10 TIME: 5:01 p.m.

We've been invaded.

Large black carpenter ants parade down the rafters of our Maine house to carpet our floor with hundreds of their cream-colored larvae and pupae. Horrified, we vacuum them all up and then leave to dine with friends. When we return after a few hours, we discover more larvae and pupae speckling our dark cherry-wood floors like grains of rice thrown after a wedding.

Here we thought our home housed just the two of us. Where are these ants coming from? Why are they traveling so far to deposit their young on our floor? Do they think *we* will tend them?

• Large black carpenter ants are common in the forest surrounding my home in Maine.

This invading horde is made up not of parents, but of sisters. All ants that tend to a nest's nursery are female, as are the wingless worker ants you see wandering about. The winged males are merely "flying sperm missiles," as biologist E. O. Wilson called them, and their sole task in life is to inseminate virgin queens (who also have wings) during short-lasting nuptial flights. When their brothers' work is done, the sisters drive these slackers out of the nest, where they will live for only another few hours to weeks. Their sisters will live on for several months. The queens, those uber-mom ants who lay all the eggs in a colony, can live for decades once they've established their colonies.

The queen births multiple broods over her lifetime, determining the sex of her progeny by controlling the flow of sperm collected during mating. She stores this sperm in an abdominal sac connected to her oviduct via a tube she opens to allow fertilization. Fertilized eggs become females, while unfertilized eggs become males. Understandably, queens fertilize most of their eggs, which hatch into industrious female worker ants. Most of these ants don't reproduce and instead tend the young, gather food for the colony, and maintain and defend the nest.

Although everyone wants to be queen for the day, like the old television show, I wouldn't aspire to be an *ant* queen. Imagine how stressful that would be: Overpopulation or other factors force you to leave your colony and all your sisters when environmental conditions are right for a nuptial flight. After mating with one or more males, you have to find a suitable spot in unfamiliar territory to create a new colony. But that's not the only challenge. You may be attacked by ants residing where you want to settle. These ants will viciously defend their nest sites and any nearby area. Because a new ant queen is also vulnerable to predation, starvation, weather, and other challenges, fewer than one in a thousand successfully creates a new colony.

• Black carpenter ants tending their pupae. All the young in an ant colony are cared for by their sisters.

Once an ant queen establishes her new abode and sheds her wings, never again to leave her nest, worker ants wait on her continually. What a trade-off! Imagine never having to work, but also never experiencing much of the world around you and becoming a mere egg-laying machine, which sounds more tedious than reading *Goodnight Moon* over and over again.

After raising her initial brood, the queen ant deploys them to tend subsequent broods. What amazing devotion these infertile sister ants have for their younger sibs! Foragers have an expandable stomach pouch that acts like a portable pantry to both process and store food they find and feed to nestmates back in the colony. Their nestmates lightly tap their antennae on the returning forager ants' mouths, prompting them to regurgitate their liquid food so the nestmates can eat it. In this way food is shared equitably in the colony, leading entomologist Thomas Eisner to claim in his book *For Love of Insects* that the food pouches of worker ants are the equivalent of a "social stomach."

In fact, studies have found that an ant will regurgitate digested food droplets at the slightest provocation—including the lightest touch of a human hair. Naturally, there are scammers eager to take advantage. Like a highway bandit, a European beetle will linger by the trail of foraging ants and literally tap their food resources. A foraging ant fooled by the beetle's tapping will disgorge a drop of liquid and hold it between its jaws while the beetle drinks. Similarly, a small fly resembling a fruit fly spends its entire life in an ant nest, where it lives off regurgitated food.

Often such trickery won't work unless the scamming insect can imitate the telltale odor of the ant species' colony. But many are indeed able to adopt the scent of another ant colony, which allows them to invade undetected. The invaders live off the colony's resources, eating their eggs, and kidnapping their young to raise as slaves in their own colony thereby increasing their workforce. Such ant slavery is common among ants in the northern temperate zones of the United States.

On a happier note, Florida black carpenter ants also have an amazing ability to moonlight as doctors. If one of their nestmates injures a limb battling other tribes, her sisters will gnaw off the injured leg to prevent infection and improve survival. The injured ant sits there, immobile, while her sisters perform the amputation.

Ants also tend to their dead. Wilson observed that a few days after an ant dies in a colony, her sisters carry her body to a dumping site outside the nest. Wilson hypothesized that the dead ant must give off a distinctive odor triggering this action. To test his hypothesis, he isolated two chemicals from dead ants and daubed them on live healthy worker ants. These very-much-alive ants were carried off to the refuse heap, kicking madly all the way, and weren't allowed to return to the nest until they licked the death scent off their bodies.

I could have used that death scent when carpenter ants besieged my house. We discovered that a leak in the roof had softened the wood in our attic, making it an attractive nesting site for the carpenter ants, who prefer to nest in dark, damp places. But after a particularly long hot spell, the nest overheated, so to save their young, the ants quickly brought them down from the attic. They had to make so many return trips to collect them all that they skipped the next step, which is to burrow out a cooler nest to lodge them in. Hence the scattering of larvae and pupae on my carpet.

But I wasn't willing to let these six-legged critters turn my living room into their next ant colony, even a temporary one. I didn't follow the advice of Wilson. As he recounted in *Tales from the Ant World*, when asked "What should I do about the [ants] in my kitchen?" he responded, "Reflect upon what you see, rather like an informal tour of a very foreign country."

"Here we thought our home housed just the two of us. Where are these ants coming from? Why are they traveling so far to deposit their young on our floor? Do they think *we* will tend them?"

I do appreciate these "foreigners," especially the important role carpenter ants play breaking down deadwood in our forest. I'm always happy to see carpenter ants *outside* my house—I just don't want them *inside*, and so we called an exterminator. His pesticides restored the house to its normal ant-free condition, but I will never forget watching those ants diligently working to save their younger siblings.

A moving instinct even I, with no sisters, could understand.

UNCOVERING AN UNUSUAL HIDING PLACE

Spittle Bugs/Froghoppers (Family Cercopoidea)

LOCATION: On a hawkweed (*Hieracium* spp.) stem in northeast Maine

DATE: June 29 TIME: 7:46 a.m.

It's early summer in Maine, so there's a chill in the morning air that requires me to bundle up in a sweatshirt and long pants. A warm brew is a must on these cold mornings. While drinking a frothy cappuccino by the bay, I see a froth of tiny bubbles bulging out from a hawkweed stem in front of me. I gently scrape away this foam and discover that it harbors a tiny, bright green spittle bug. This bitty critter is so vulnerable to the elements, so in danger of being eaten by voracious flying giants or attacked by stinging monsters, that it builds a castle and hides inside it day and night—a castle made of bubbles.

• This foam, often seen on plant stems, is made by a spittle bug to protect it while it's young and vulnerable.

• This froghopper is less than half an inch long, but it can jump close to a yard.

Spittle bugs are the juvenile form of froghoppers. These omnipresent froghoppers, cousins to aphids, are no bigger than your pinky fingernail and, if you look close enough, have comically bulbous noses and big eyes. Good luck getting close enough, though—as soon as you come near a froghopper, like its namesake, it jumps away. But when it first emerges from its egg as a spittle bug, this grain-sized baby insect doesn't yet have its hopping chops and must protect itself by other means.

That's where the bubbles come in.

The spittle bug sucks on plant sap and then pees out a cocoon of bubbles that will protect it from drying out, overheating or freezing, or being eaten by birds, spiders, and various insects, none of whom like the bubbles' acrid taste. The foam also prevents the bug from being fodder for larvae hatching from eggs laid by wasps, who can't get a strong enough footing on the bubbles to insert their eggs into the spittle bug's body.

Eventually, like a baby outgrowing its diapers, the spittle bug grows big enough to transform into a froghopper, leaving behind its castle of pee. This froghopper sports a drab coat of mottled brown and tan. It no longer needs its bubble castle to hide in because it now has long wings to cover its body, protecting it from the elements. It also has a new way to escape predators: hopping.

And boy, does it hop! The froghopper is less than half an inch long. But it can jump close to a yard, the equivalent of a human jumping the length of a football field. Attached to the froghopper's hindlegs are bendable straplike structures that the bug keeps flexed and pulled back like the elastic bands of slingshots. The instant it finds itself face-to-face with the gaping maw of a robin or the enormous face of a human, it releases these straps, which catapult the bug forward a distance a hundred times the length of its body. An amazing feat that must be fun too. I imagine these little bugs sucking those stems and then, when predators creep close, having a good excuse to leave the dinner table, dumbfounding an expectant bird or wasp by soaring through the air like a trapeze artist in a circus. *Hate to eat and hop but I gotta go.*

• A meadow froghopper, the adult form of a meadow spittle bug.

DISCERNING THE CHORUS WE CAN'T HEAR

Candy-Striped Leafhoppers (*Graphocephala coccinea*)

LOCATION: On a milkweed (*Asclepias*) leaf in a garden, northeast Maine

DATE: September 22 TIME: 2:16 p.m.

It's a Sunday in July when the lobstermen don't work, so the bay is quieter than usual. The sweet trill of a warbler calls my attention, but despite my efforts to see it with my binoculars, it remains hidden in the greenery of a maple tree. So I take a stroll in the garden to see if I can find some critters closer to the ground instead.

It would have been easy to overlook the first insect I find, as it measures in mere millimeters, but what a feast for the eyes this leafhopper becomes once I use my camera lens to magnify it on a milkweed plant. Its sleek, neon-

• This candy-striped leafhopper communicates with other leafhoppers by creating vibrations that travel through plant stems and leaves.

bright turquoise-and-scarlet converging stripes are reminiscent of those of a race car, and another black racing stripe linking its black eyes contrasts with its canary-yellow, cone-shaped head. It's more beautiful than the flowers of the milkweed whose leaves it's sucking.

The leafhopper's name comes from the fact that, to escape predators, it can leap forty times its own body length, making it a champion jumper of the animal world. Despite those relatively big hops, leafhoppers like the one I captured with my camera can stay on a single plant, if not the same branch or leaf, for most of its entire monthslong lifetime. There, it feeds itself by sucking out the plant's juices. The leafhopper will molt several times until it matures enough to mate. The mated female lays her eggs in the plant's stems and leaves, repeating the life cycle.

But finding a mate can be challenging when you're so tiny and don't venture far during mating season. How can you produce a sound big enough to be detected across a distance? The airborne sounds we hear don't usually travel very far unless they are very loud because they spread outward in all directions from their source and thus dissipate with distance. A leafhopper is smaller than the size of your pinky's fingernail—too small to produce a sound loud enough to travel the distance to potential mates.

Instead of making vibrational signals in the air, like those we hear, leafhoppers and many other kinds of insects use muscles in their chest and abdomen to create vibrations that others of their species will *feel* rather than hear. The vibrations travel down their legs and through plant surfaces, triggering sensory hairs on the legs of other leafhoppers on the same plant. Because these vibrations move along the more direct paths of branches and leaves, they retain energy over longer distances than soundwaves dispersed in air, and thus they can be more complex and lower in frequency, which we perceive as pitch, than the monotonous chirping of crickets or buzzing of cicadas. How do we know this if the vibrations don't make a sound loud enough for us to hear? Scientists use devices to amplify and record their vibrations, or they use lasers to detect their vibrational waves and then translate their frequencies into something audible to human ears.

When I learn this, I immediately jump onto the internet to hear these sounds myself and am blown away by how many of them remind me of those made by much bigger animals—specifically, whales—as they slowly increase in pitch all while staying in a low bass register. Others sound more percussive, with repeated segments, or like the hooting of an owl, the mooing of a cow, or the sound of a song on a cassette tape played backward slowly, if you're old enough to remember what that sounds like. As the narrator of *bioGraphic* magazine's "Invisible Nature: Code of the Treehopper" video essay noted, "These vibrations are happening all around us. When we listen in a new way, a whole world is opened up—a world that's been there all along."

That's because close to 200,000 species of insects, including treehoppers, cicadas, crickets, and katydids, are thought to use these surface vibrations not just to find mates but to send other messages, like "Watch out for other animals!" or "Let's gather together as a group." Babies use the vibrations to call their mothers, and mothers vibrate back a silencing command so predators won't detect them. I wish I could be like Horton in *Horton Hears a Who!* and eavesdrop on all these vibrations. There must be a lot of chatter going on in my garden, beyond all the buzzing of the bees and chirping of crickets. "A single stem might be as raucous as a busy street, full of cries for help, calls for silence, invitations to hang out, and literal booty calls," Ed Yong mused in his book *An Immense World*.

Here I thought it was a relatively quiet morning, but if I had been able to tune in to a different register, I would have heard a cacophony, all coming from tiny beings compelled, just like us, to convey their deepest fears, needs, and desires.

A chorus we can't hear.

TRACKING THE GREAT MIGRATION

Leafhoppers (Family Cicadellidae) and Many Other Insects

DATE: August 15 TIME: 2:47 p.m. (leafhopper)

While taking a "Bugs in Winter" class, I read that to escape the winter cold and paucity of food linked to this frigid season, certain leafhoppers (like the candy-striped leafhopper described in the preceding chapter) migrate thousands of miles to their wintering grounds in the southern United States. How can creatures no bigger than a grain of rice fly so far?

They don't—or at least not without the help of the wind.

Instead, leafhoppers and many other insects big and small hitch a ride on that insect superhighway in the sky, nearly as high as the clouds. More than seventy different insect species in North America travel great distances south in fall and north in spring with the aid of weather fronts. In what

• Leafhoppers, like this one, and many other insects travels thousands of miles via multiple generations each year.

is described as a vacuum cleaner effect (and I imagine is kind of like the tornado that transports Dorothy to Oz), the drop in atmospheric pressure of weather fronts sucks insects high up into the aeolian zone, named after the Greek wind god Aeolus. Some insects are swept up by accident, while others, by positioning themselves properly on vegetation late in the day when winds pick up, intentionally surf these air currents to get to other places.

Consider the potato leafhopper (*Empoasca fabae*), which hitches a ride on these reliable winds in the fall, traveling from the Great Lakes to the Gulf Coast like a Midwestern snowbird. In spring, after bulking up on the fats and sugars found in pine needles, the potato leafhopper once again rides the winds, this time heading back north. Once it arrives in its northern grounds, where food is more plentiful, it starts reproducing, often producing several generations during a summer. The last generation of the summer season heads south. In this way the leafhopper completes its spectacular annual circular migration involving thousands of miles and multiple broods of the bugs—a multigenerational round trip.

And it's not just leafhoppers that enter the superhighway in the sky. As researchers continue to ply the aeolian zone, the list of insects making such round-trip seasonal migrations riding the wind continues to expand and includes a number of those I had become acquainted with in my garden or nearby marshes, including monarch, American lady, painted lady, red admiral, cabbage white, and sulfur butterflies, as well as skimmer dragonflies, grasshoppers, milkweed bugs, green darners, and lady beetles (ladybugs). Researchers once detected a lady beetle some 6,000 feet above the ground! Bigger insects can tolerate the greater wind speeds found even higher above the earth than where the leafhopper commutes, and some of them are able to travel more than 400 miles per night on that wind. Most insects traveling the superhighway in the sky still have to keep flapping their wings to stay aloft, but without the wind they couldn't go so far.

The total distances some of these insects migrate are epic. Painted lady butterflies each spring rise from tropical regions of Africa, cross the Sahara Desert and Mediterranean Sea, continue through Europe, and land as far north as the Arctic Circle. These *Vanessa* species (a nod to their genus name—I like to be on a first-name basis with all my insects) remain on top of the globe for just a few weeks before their offspring start the reverse route south in the fall, a round trip involving up to six successive generations and a total of 9,000 miles.

The number of insects taking advantage of these aeolian winds is also epic. Researchers in Great Britain and Israel estimate that 3.5 trillion insects migrate above the southern United Kingdom each year, far outnumbering the 2 billion migrating birds thought to traverse the same swathe of sky each year. Insect migration "represents the most important animal movement in terrestrial ecosystems," wrote the researchers in their 2016 report in *Science* magazine. Millions to billions of insects, some migrating and some accidentally swept up by winds, can fill the aeolian airspace at any one time. In the 1920s, entomologists "estimated that at any given time on any given day throughout the year, the air column rising from 50 to 14,000 feet above one square mile of Louisiana countryside contained an average of 25 million insects and perhaps as many as 36 million," Hugh Raffles explained in *Insectopedia*.

For many insects, such a risky journey can also be a life-ending one. They can starve to death, get eaten by fellow travelers, or be wind-battered into oblivion. But the freezing winter is also a life-ender. So after spending most of their lives never tarrying farther than plants rooted to the ground, nature prompts them to take that leap off the leaf—to rise upward in search of something better, the wind helping their wings carry them to other lands.

FOLLOWING FALL FAIRIES

Woolly Aphids (Family Eriosomatinae)

LOCATION: Flying above a garden in northeast Maine

DATE: November 11 TIME: 1:13 p.m.

While putting my Maine garden to bed for the winter, I take time out to marvel at tiny insects suddenly omnipresent in the air and sparkling like dust motes in the sun. Saffron and scarlet tree leaves carpet the ground and are feathered by the morning frost. What could still be flying about so long after the butterflies and bees have left us?

Unlike swarms of frenetic darting gnats, these luminescent specters are more spread out and move more slowly. They seem to drift on the breeze and are so abundant this fall that the Facebook pages of the Maine Naturalist and Maine Insects groups are abuzz with aphid sightings—commenters describe them as floating fluff or lint, snow flies, and blue fuzzy butts (for

• I used to think of aphids as always latched on to the plant stems they suck, but woolly aphids, like this one, have a flying adult form that can be abundant in fall.

• Black carpenter ants tending to woolly aphids.

their diaphanous powder-blue tutus), and some mistake them for plant seeds dispersed by the wind.

Fortunately, these little buggers aren't skittish, and when one lands on my palm I take a picture and submit it to iNaturalist. Experts identify it as the adult form of a woolly aphid. I am already well acquainted with aphids, often spotting them sucking on the stems and leaves of my garden plants, but who knew these insects could also wing it? And wing it they do, having the final fall fling of insects destined to spend winter as eggs tucked into tree bark crevices, assuming chickadees and other birds staying up north for the winter don't eat them.

At first, I don't connect these delightful insects to a more prosaic colony of woolly aphids I had seen earlier in the summer at a nature preserve. Those immature aphids were sooty-colored and rough-textured in a regular (tessellated) fashion, like miniature hand grenades with white feathery streamers attached to them. Those streamers are from the wax they secrete to make them distasteful to predators and to also possibly help them stay aloft when they become winged adults.

Zooming into the photo I'd taken of those juvenile woolly aphids, I could see some lighter newborn aphids hiding under their larger and woollier mothers. Unlike most insects, multiple generations of aphids are born live from virgin females instead of from eggs—and in quick succession. These cloned aphids can rapidly populate an area when food is abundant, so by the time an aphid has eaten her way up a plant stem, she's become a great-great-grandmother. Mating and fertilization, with its more environmentally adaptive gene swapping, is saved for fall.

Carpenter ants had been minding the wingless aphids I saw at the preserve. These ants protect them from predators and herd them to new host plants when sap becomes less nutritious or abundant. In exchange, the ants dine on the aphids' sugary excretions, called honeydew. They "milk" aphids by gently tapping their abdomens with their antennae, causing them to expel a honeydew droplet. Studies show that such "farmed" aphids produce more honeydew in their lifetimes than those not tended by ants.

There are hundreds of species of woolly aphids, most of whom are picky eaters; some preferring to dine solely on beech, birch, or apple trees, for example, while others prefer alders.

Some adult woolly aphids wait out winter on tree branches in tight clusters encased in their own waxy wool. But shorter daylight hours trigger most of them to produce their final flying forms. These glitter the nippy air during sunlit fall days as they try out their new wings while looking for mates and a suitable tree on which to lodge their eggs. Unlike their predecessors, many of these aphids get lucky and have sex before they die.

But first they put on a dazzling display.

• A woolly aphid rests on my finger.

APPRECIATING WASPS, PART 1

Great Black Digger Wasps (*Sphex pensylvanicus*)

LOCATION: On a summer sweet (*Clethra*) bush in a garden, northeast Maine

DATE: August 16 TIME: 5:11 p.m.

I can't believe I'm inches away from a huge wasp and trying to take its picture.

I've been terrified of wasps ever since one of the girls at my best friend Judy's tenth birthday party got stung by one. She screamed, and the rest of us jammed ourselves into the corner, cowering, clutching our sleeping bags, as the giant wasp flew about with its long black legs dangling down, until Judy's dad finally captured it.

I know I'm not the only one with a distaste for wasps. As behavioral ecologist Seirian Sumner wrote in her book *Endless Forms: The Secret World of Wasps*, "Wasps are depicted as the gangsters of the insect world; winged

• Although I was once afraid of wasps, I was mesmerized by the beauty of this great black digger wasp on a summer sweet bush.

thugs; inspiration for horror movies; the 'sting' in the tale of thriller novels; conduits of biblical punishment. Shakespeare, Pope Francis, Aristotle, even Darwin struggled to speak favorably of wasps, and questioned the purpose of their existence."

So what am I doing so close to this wasp, whose sleek dark body extends longer than my thumb and certainly bears a powerful stinger?

Admiring her beauty and wondering why she's drinking from my flowers.

No question about it—she's mesmerizingly gorgeous, with smoky black wings shimmering a cobalt blue iridescence above her ebony body. She outshines the dull white summer sweet flowers she plies. But what's she doing in the garden, behaving more like a vegetarian bumble bee than a blood-thirsty carnivorous wasp?

Most wasps, like this great black digger wasp, are solitary souls, and only the females sting, mainly to paralyze the high-protein insect prey they feed their progeny. Males are deadbeat dads whose sole contribution to the next generation is their sperm. Unlike her young, the single expectant wasp mom dines on flower nectar and pollen and wouldn't waste her venom on a person unless she thought her life or her eggs were in danger. Yellow jackets and other social wasps are more aggressive and likely to sting people to protect their large nests. One researcher (much braver than I) had the audacity to repeatedly snatch katydids from the clutching legs of the great black wasps that caught them. The wasps tried hard to get the insects back, but they never resorted to stinging the guy. A great black wasp will do anything for her babies, even though she won't usually live long enough to see them hatch. Most of the great black wasp's heroic parenting occurs *before* her babies are born.

Evolution has honed the female wasp into a fantastic creature with amazing instincts, and like her bee cousins, she's an incredible learning machine that can adapt quickly to changing circumstances. The needlelike

• The tiny, twisting waist of this thread-waisted digger wasp lets her precisely position her stinger to paralyze prey.

extension (ovipositor) she uses to lay her eggs evolved into a stinger that can inject paralyzing venom into prey. French naturalist Jean-Henri Fabre called this wasp stinger a "mother's stiletto," probably referring to the stabbing weapon of medieval assassins (I prefer to imagine it as the glamorous shoe).

Along with their stilettos, digger wasps also adopted another new style—tiny, twisting waists that make them nimble in nabbing and paralyzing insects with their precisely positioned stingers. Why paralyze and not kill prey outright? Because these insects have to remain fresh weeks later in order to serve as food for the wasp's hatching babies, given that no parent will be around then to feed them.

To protect her prey and the grubs that will feed on it from predators and weather, the mother wasp burrows out a nest in the ground. Her biting jaws and rakelike legs loosen and scrape away soil to create a multichambered nest. She outfits each chamber with an egg and a bunch of prey insects, which she immobilizes with just the right amount of venom to permanently paralyze them. She's an excellent neurosurgeon, administering her venom with astonishing precision. In the large prey the great black wasp specializes in catching, she uses her flexible waist to precisely place her stinger in each of the insect's three nerve centers (ganglions). That way her venom can immediately stop it from moving its legs and wings.

Digger wasps are mighty particular about their food. The great black will prey only on grasshoppers, locusts, and katydids. Another digger wasp species hunts not just any spiders but specifically *pregnant* spiders. That way the wasp's young can feed on their abundant eggs.

Nineteenth-century French naturalist Léon Dufour noted that in twenty years of hunting for a particular type of jewel beetle aboveground, he found only one, whereas he unearthed hundreds of these beautiful beetles in digger wasp burrows. This type of beetle lives mostly underground or burrowed inside trees. How did a wasp find them? With her antennae, which have both odor and vibration sensors and "are the insect's equivalent to a metal detector, revealing hidden secrets of the underworld," Sumner wrote, although she noted that these sensors usually detect an insect only if it is shallowly buried in soil or wood. The great black wasp can more easily detect the scent trails of her prey that are out and about because their odors spread far and wide.

Some digger wasps make excellent housekeepers. To make sure prey and eggs don't decompose or dry out, the beewolf digger wasp encases them with a water-repellent substance. She then uses her antennae to insert antibiotic-producing bacteria into each baby burrow. And her eggs exude an antifungal gas commonly used by farmers to fumigate their fruit crops. "The magic combination of fumigation and embalming provides beewolf babies with the cleanest nursery nature can buy," noted Sumner. And Dufour observed, "How immensely superior to our own pickling processes is that of the wasp. What lessons can we not learn from her transcendental chemistry!" Just as ants invented farming before we did, wasps discovered (and used) antibiotics and embalming fluid before us. If they were some alien species equal to our size arriving on Earth from another planet, instead of miniscule insects flitting about the corners of our lives, we would think them brilliant!

As a final act of her pay-it-forward parenting, the digger wasp fills up her burrowed tunnel with the mound of expelled soil at the nest entrance, sometimes biting off more soil around its opening if needed. She then compacts this soil so it's indistinguishable from surrounding ground. She often uses tiny tools, like pebbles, leaves, or small sticks, pushing them with her head to press down the soil, showing once again that insects mastered something way before we did—tools.

The digger wasp will modify her housekeeping routine when prodded by pesky experimenters, suggesting she's not operating by instinct alone. Fabre observed that, prior to bringing her prey into the prepared chamber, one type of digger wasp routinely dropped the paralyzed insect beside the nest entrance so she could make a final inspection of the nursery. When Fabre moved the prey a few inches away while the wasp did her look-see inside the

• A great black digger wasp with her prey. Digger wasps are very choosy about the prey they paralyze and feed to their young.

nest, she emerged from it seemingly surprised not to see her catch beside its opening. She would then find the paralyzed insect and drag it back to the nest entrance before beginning her inspection routine yet again. This left her prey unguarded, so that Fabre could again move it away from the burrow entrance. After Fabre played this trick two or three times, some of these digger wasps caught on and adjusted their behavior accordingly, pulling their prey into the chamber without examining the nursery again, showing Fabre they did have some wits about them.

The great black wasp also must have good spatial memory in order to be able to find her nest after making hunting excursions as far as seven miles away. As Lars Chittka noted in *The Mind of a Bee*, "Evolution won't be forgiving of a mother who forgets the location of her own home that contains her offspring." Since other digger wasps are likely to have made their nests nearby, she must remember the unique landmarks and odors of her own nest. Supporting this notion, the brains of wasps constructing nests and provisioning their broods have the same sort of elaborate convolutions seen in mammal brains. And compared to their more primitive, homeless relatives, nest-making wasps have an enlarged portion of the brain where memories reside. Plus, studies have revealed that wasps can hold the memory of a spatial map in their bitty heads for at least twenty-four hours.

So please give the great black wasps, cicada-killer wasps, and other solitary digger wasps commonly found in our own backyards the respect due to them and try not to be afraid of their stings, which they rarely waste on humans. Think of these impressively large wasps as pest controllers and pollinators instead of pain-inducers.

Because of their great and highly specific hunting skills, companies have put solitary hunting wasps to use as biocontrol agents for cockroaches and coffee berry borer beetles. Some experts consider the great black wasp a friend to home gardeners as well, since a single female can capture over a dozen plant-eating insects a day. The wasp is known for pollinating plants in the milkweed, carrot, bean, and other families. "In her mission to feed herself and her larvae, a female great black wasp will clean up your garden of herbivorous pests and pollinate flowers into the early fall," explained Paige Waddick for the University of Minnesota Department of Entomology's profile of the species.

Even the Orkin pest control company has good things to say about the great black wasp, writing on their webpage: "Since the great black wasp is not aggressive and is an important predator of harmful insects and a good pollinator of flowering plants, there is no reason for the homeowner to control them." Sumner recommended that we "admire them for their . . . curious life stories and as marvelous playthings of evolution." When we see one, she said, we should "pause, smile and give it an appreciative nod, thanking it for reminding us of the many hidden charms and wonders on the planet."

And let's not forget that the great black wasp is breathtakingly beautiful. When I take a break from working on this chapter and open my front door, I discover one of these gorgeous beings on my doorstep. She captivates me once again with her brilliant blue iridescent wings before flying away. I wish her luck in all she must do in her short life for the next generation, knowing she will never take the time to smell the roses unless they harbor her prey. I hope we meet up again in the garden as summer wanes into fall.

• A digger wasp flies to her nest clutching a paralyzed weevil to feed her young.

APPRECIATING WASPS, PART 2

Family Ichneumonidae

LOCATION: On the glass door of a home, northeast Maine

DATE: August 27 TIME: 9:25 p.m.

It resembles a smoky nugget of dark amber with a protruding white braid running down its middle. Amber starts off as tree resin that hardens over millennia, but this inch-long egg case clinging to my flowerpot in the heart of Philadelphia started off as a praying mantis's frothy secretion that hardened within a few hours to protect the hundreds of eggs within it through winter. Every day I check the egg case, eager to see baby praying mantises emerge. It tantalizes like a present waiting under the Christmas tree.

The weather warms but there is still no sign of mantises. By now the eggs should have hatched, the baby mantises exiting through the overlapping scales in the case's center braid. Another week goes by and I still don't see the

• This wasp (*Opheltes glaucopterus*) resting on my glass door parasitizes immature sawflies.

tiny hatchlings in my garden. Since baby mantises can't fly when they first emerge, I figure I'll spot them on my plants. After more days pass with no signs of mantises, I look more closely at the egg case and notice several tiny, perfectly round holes.

When I hop on the internet to find out what these holes might be, I discover they are likely the exit holes made by parasitic wasp larvae that ate up the mantis eggs and then skedaddled. That's the moment when I first become aware of parasitic wasps, a huge category of insects thought to far outnumber any other group of organisms on the planet. Genetic studies done by biologist Andrew Forbes of the University of Iowa recently determined that one common parasitic wasp species, first identified in the 1840s, is actually sixteen different species. And still more are being uncovered. Experts estimate they've only identified about a tenth of the members in just one family of parasitic wasps. "The bottom line is there are astronomical numbers of parasitic wasp species [out] there—you find more species every time you go out," said entomologist Andy Austin in a 2020 interview posted on YouTube by the South Australian Museum.

That's no surprise, since experts suspect that every species of insect is parasitized by at least one species of wasp, if not several. Even insects living in aquatic environments can't escape them. Plus, insect species may be targeted by different parasitic wasps at different stages of their life cycle—one attacks the eggs, another the larvae, a third the pupae, and a fourth the adults. This begs the obvious question: If these apparently infinite species of parasitizing wasps are so common, why don't we see them?

Because so many are tiny and hide inside the bodies of others.

Hundreds of thousands of wasp species exploit other insects, but most are no bigger than the eye of a needle. For some, including those parasitizing parasitic wasps—now that's a tongue twister—you'd need a microscope to see them. The smallest, aptly called fairy wasps, are no bigger than a single-celled amoeba yet have all the features of a bigger wasp—incredibly miniature legs, antennae, and mouths.

One entomologist found a series of four parasitizing wasps on a single caterpillar like a bunch of nesting Russian dolls—one ichneumonid wasp species parasitized the moth caterpillar, but then this wasp was parasitized by another ichneumonid wasp species. Not to be outdone, two wasps from a different wasp family parasitized both parasitizing ichneumonid wasps on the caterpillar.

I'm blown away by the endless array and immense sway of tiny organisms I didn't even know existed a day ago. How do these insects live?

I discover that parasitic wasps are not anything like their cousins, the solitary digger wasps we explored in the previous chapter. Neither does their behavior resemble that of the social yellow jacket wasps, diligent parents who, over a period of weeks, seek out and chew up wood and other plant fibers and then painstakingly dab together the macerated pulp to build elegant basketball-sized hives in which to rear their progeny. Worker wasps feed these young with bits of insects and carrion they've thoughtfully prechewed.

Parasitic wasps shirk such responsibility and instead rely on other insects to feed and/or house and protect their progeny. Their needlelike ovipositors are adept at piercing everything from the soft bodies of caterpillars to the hard shells of the praying mantis egg cases they inject their eggs into. The larvae then hatch and ungratefully eat their hosts.

Freeloading is a way of life for these wasps. "Their lives are grisly and sinister, but their abilities are incredible," commented Ed Yong in an April 2018 issue of *The Atlantic*. "It's easy to see parasites as . . . unsavory creatures, but they're actually practicing the most successful lifestyles around. The defining innovation of the animal kingdom is not the stone tool of the ape nor the flight-capable feather of the bird, nor the hive mind of the ant, but the egg-laying stinger of the parasitic wasp."

Wasps sponge off as many as half of all caterpillars, killing them in the process—a valuable form of insect control. Some farmers deploy parasitic wasps like paratroopers to destroy caterpillars before they ruin their crops. Many parasitic wasps inject a paralyzing substance into caterpillars and then bury them in the ground or in other cavities before injecting their eggs into the immobilized insects. When their young hatch, they munch on the insides of the fresh caterpillar *while it is still alive.*

I'm both mesmerized and grossed out by a YouTube video showing a wasp laying close to fifty eggs inside the caterpillar of a white cabbage moth. The camera zooms in on the glistening wasp grubs, who, after digesting much of the caterpillar's unessential tissues, use their sharp chitinous teeth to cut through the insect's tough skin and then wriggle their way out, while ominous music plays in the background, reminiscent of something out of the movie *Alien*. The narrator, who has a charming British accent, reveals that, remarkably, not a drop of blood has been spilled in this process and that, unbelievably, the caterpillar survives this massacre.

In a case of serious Stockholm syndrome, this caterpillar then sticks around to fiercely defend the wasp grubs. As the grubs busily spin little golden cocoons around themselves so they can morph into their adult forms and take flight, the ravaged caterpillar gets up on its hind legs and swings its green bristled body to and fro to defend itself and its wasp inhabitants from predators. It also makes its own silk to help cover the pupae and keep them safe while they grow wings. By the time these wasps fly out of their cocoons, their poor, sad-sack sacrificial host caterpillar will be dead from starvation.

The white cabbage moth caterpillar normally doesn't show such protective behavior even toward the young of its own species. The reason it does so now is that the parasitic wasp has taken over not only its body but also its mind. How it performs this trick remains a mystery, but it's not unusual. Research reveals that other parasitic wasps also cause bizarre

• The long, needlelike ovipositor of this parasitic wasp, *Rhyssa lineolata*, is used to reach, paralyze, and then lay eggs on the larvae of insects buried deep in wood.

behavior in their hosts. There's one that injects its venom into the brain of a cockroach to turn the insect into a zombie. This enables the wasp, using her latched-on antenna like a leash, to walk the cockroach back to her nest, where it will become fodder for her young.

Cabbage moth caterpillars have a primitive defense that could protect them from wasp invasion, one that propels their blood cells to encapsulate implanted eggs or young larvae and smother them to death. But the wasp circumvents these defenses by inserting, along with her eggs, viruses that disarm the caterpillar's immune system as well as tweak its development to favor the growth of the wasp grubs. Over tens of millions of years of evolution, those viruses have become so entwined with wasp eggs that they can't spread without the wasp. And they're very picky. Because each virus species can infect only one caterpillar species, hampering its immune responses, the wasp harboring the virus can also afflict only this species.

How does this parasitic wasp find its very particular host for its young among a host of other species? With its astonishing oversized antennae, which are equipped with more than twenty chemical sensors. When a white cabbage moth caterpillar chews on its own host plant, the leaves emit distinctive chemicals (scents) that waft in the air for quite a distance and are picked up like radio signals by the wasp's antennae. Homing in on those signals brings the wasp to the plant where the caterpillar lies. Then, the volatile chemicals given off by the caterpillar, whether through its saliva, droppings, or silk, enable the parasitic wasp to zero in on the caterpillar itself.

Parasitic wasps bothered and bewildered Charles Darwin, who went to a theological seminary before studying natural history. "I cannot persuade myself that a beneficent and omnipotent God would have designedly created the Ichneumonidae [a family of parasitic wasps] with the express intention of their feeding within the living bodies of caterpillars," he wrote in a letter to American botanist Asa Gray in May 1860. On the other hand, referring to his emerging theory of evolution, Darwin added, "I am inclined to look at everything as resulting from designed laws, with the details, whether good or bad, left to the working out of what we may call chance. Not that this notion *at all* satisfies me. I feel most deeply that the whole subject is too profound for the human intellect. A dog might as well speculate on the mind of Newton."

It's easy to see the parasitic wasp as "bad" and the caterpillar as "good" or at least the victim. But nature makes no moral judgments. We've seen that social bees and wasps have evolved cooperative and altruistic behaviors because they benefit their own colonies. And studies by the late animal behaviorist Frans de Waal and others show that some social animals, such as primates and elephants, have evolved the two basic pillars of morality—fairness and empathy—because they benefit their own tribes. But the solitary parasitic wasp injecting her eggs into another species of insect has different evolutionary pressures shaping her behavior. She carries out what millions of years of unintentional chance events have shaped her to do with awe-inspiring effectiveness. Can we ignore the plight of the caterpillar and instead see the complex maneuverings of the parasitic wasp as a polished act worth applauding, one to revere rather than reject?

Instead of asking why they exhibit such gruesome behavior, maybe we should be asking, why not? Anything's possible in the world of nature. Whether the parasitic wasp's progeny carries on her lineage at the expense of the caterpillar or praying mantis boils down to impersonal chance, to the luck of the draw. This concept competes with the morality that we humans are so proud of. But Taoists believe that distinctions between good and bad, and the moral judgments attached to them, are a matter of perception. Both are thought to be part of an indivisible whole—one can't exist without its complement. Rabbinic thought also promotes the notion that good and evil must coexist but claims they are in continual battle within individuals, something Soviet dissident Alexandr Solzhenitsyn seemed to reference in *The Gulag Archipelago*, in which he wrote, "The line dividing good and evil cuts through the heart of every human being. And who is willing to destroy a piece of his own heart?" If we view nature as a whole, we could take Solzhenitsyn's question to heart and ask, "Who is willing to destroy (or look askance upon) a piece of their own world?"

APPRECIATING WASPS, PART 3

Cuckoo Wasps (*Chrysis* spp.) and Potter Wasps (*Ancistrocerus* spp.)

LOCATION: Inside and around a lawn chair beside a bay, northeast Maine

DATE: August 26 TIME: 8:05 a.m.

I'm sitting by the bay in Maine, enjoying the cool breeze coming off the water and trying to read, but a tiny flying emerald jewel keeps distracting me. I see it land on the chair next to me and crawl around on its armrest support—and then it disappears.

What is it and where did it go?

Eager to take a picture of this beauty, I keep watching, and eventually it flies by me again. This time I track it closely, watching it land on the chair again and then vanish into a hole in the chrome armrest. On further

• A cuckoo wasp just before invading and laying her eggs in the nest of a potter wasp, who provisions them.

inspection, I see that the hole is partially filled with mud—I suspect it's a mud nest made by a wasp or bee to stash their eggs and food for the larvae to eat once they hatch.

I return to my book but am distracted yet again, this time by a large ebony insect flying by, with a distinctive narrow waist, long and narrow amber-tinted wings held on either side of her back, and an abdomen ringed with yellow—telltale signs of a wasp. The wasp has a bit of mud in her mouth, and she too disappears into the same hole to which the emerald insect was drawn. But a wasp wouldn't share her nest with another species, right? So what's happening here?

I solve the mystery by identifying the two insects. The larger is a potter wasp, so named because of the mud nests she creates, which sometime resemble pottery vessels. The iridescent insect, which is chubbier than a typical wasp, is a cuckoo wasp.

The cuckoo wasp is so called because, like the cuckoo bird, it lays eggs in the nests of other species—usually those of other solitary wasps or bees. When the larvae of this cuckoo wasp hatches, they will gobble up the eggs and larvae of the potter wasp, as well as the provisions the potter wasp carefully collected for its young and placed in the mud cells. I look inside the hole and see both wasps. You'd think the potter wasp would kick out the murderous cuckoo wasp, but these cuckoos can mimic the smell of other species, and the potter wasp is none the wiser.

I want to chastise the cuckoo wasp for being such a slacker mom—and for killing off the young of another species. But then again, as the proverb goes, it takes a village to raise a child. As we saw in "Trailing the Very Busy Bumble Bee" (page 1), bees know this principle well, with their large groups of workers who together raise the young in the hive. The cuckoo wasp seems to know that she, on her own, is not up for this task. Maybe she'd feel differently if she had a mate that helped out?

• A potter wasp making her mud nest in my lawn chair.

I didn't need a whole colony to help me raise my young, but I could have used the help of another person who could be around during the day while my husband worked. I especially had a difficult time during the newborn stage, finding it exhausting and lonely. Seeing the world through my baby's eyes rewarded me with renewed wonder, and I felt an outpouring of love for the vulnerable creature I held in my arms. But I also found tending an infant to be extremely demanding and literally belittling by leaving so little room for me.

At a family education session on Passover, the holiday that celebrates the Jewish Exodus from slavery in Egypt, my rabbi asked the group to think of new ways to interpret slavery, ways in which we ourselves might be enslaved and needed to be set free. I shared that I felt a feeling of enslavement while tending a newborn through all hours of the night and day, never getting any

"I want to chastise the cuckoo wasp for being such a slacker mom—and for killing off the young of another species. But then again, as the proverb goes, it takes a village to raise a child."

time to myself or much needed sleep. The rabbi was horrified, as if I had confessed that I wanted to murder my child. I felt chastened and questioned whether I lacked the maternal instinct I thought all other women seemed to have.

Recently, this sacred notion of a maternal instinct has been called into to question. As Chelsea Conaboy, author of *Mother Brain: How Neuroscience Is Rewriting the Story of Parenthood*, noted in a *New York Times* editorial:

> Darwin codified biblical notions of the inferiority of women and reaffirmed the idea that their primary function is to bear and care for children.
>
> "What a strong feeling of inward satisfaction must impel a bird, so full of activity, to brood day after day over her eggs," Darwin wrote in *The Descent of Man, and Selection in Relation to Sex* in 1871. Observant as he was, Darwin apparently ignored the hunger of the mother bird and the angst of having mouths to feed and predators to fend off. He didn't notice her wasting where wing meets body, from her own unending stillness.

Conaboy stressed that caring for children is not merely a hormonally fueled instinct but a learned behavior as well, one that both mother and father, or adoptive mother and adoptive father, develop over time in response to caring for a child. Brain imaging studies in humans and extensive animal research reveal that the parental brain takes time to come into being, driven as much by exposure to the powerful stimuli babies provide as by the hormonal shifts of pregnancy and childbirth.

There's no denying that caring for young children can be stressful, especially if you bear the full brunt of their care. In a 2024 advisory, even US Surgeon General Vivek Murthy felt compelled to write:

> Parents who feel pushed to the brink deserve more than platitudes. They need tangible support. . . . Many parents and caregivers I've met say it's not easy to ask for help when everyone is grappling with hectic schedules and when it feels as if other parents have it all figured out. As hard as it is, we must learn to view asking for help and accepting help as acts of strength, not weakness. . . . Parenting at its best is a team sport.

I adapted to my 24/7 job of tending to a baby by seeking out more help from my husband and others, eventually hiring a part-time babysitter so I could have a few hours to myself each day to catch up on my sleep, write, and refuel mentally and emotionally. Animals like cuckoo birds and wasps have developed clever but crueler survival strategies to rely on others to tend their progeny. I still feel sorry for that potter wasp, who went to all that effort to support her young only to have them devoured by another species. But nature is more rational than just, shaped by chance, not by moral design, so I have to stop judging her.

And myself.

EXPOSING INSECTS WITH A LOT OF GALL

Rose Gall Wasps (*Diplolepis* spp.) and Other Gall-Making Insects

LOCATION: Inside the gall of a wild rose (*Rosa* spp.) bush beside a bay, northeast Maine

DATE: November 11 TIME: 9:46 a.m.

One November morning, while admiring the orange and rusty red hues autumn fairies painted on a wild rose bush clinging to the shore of our bay, I spot something odd—spherical structures mottled in shades of beige anad brown and covered with menacing long spines. These aren't the ordinary rose hip fruits that ripen at the base of each rose flower after its petals fall off. Rose hips resemble miniature squat pomegranates; they turn scarlet in fall, their smooth skin encasing abundant seeds that you have to spit out in order to sample the sweet-tasting paste surrounding them. No, what I'm looking at are not at all inviting to the

• Many insects induce plants to grow galls, such as the spiny structures seen on this rosebush. The gall feeds and protects the developing insect inside.

mouth. They resemble the spiked balls medieval militaries flailed against their enemies. What are they and who made them?

I suspect they're galls with insects hiding inside them, having recently taken Charley Eiseman's field study class on signs of insects (see "Reading Insect Lore in Leaves," page 129). The most likely culprit is a wasp or fly. Galls form on plants in response to insect, mite, or microbial intruders.

Galls can appear anywhere on a plant, from root to shoot, leaf to flower to seed, or trunk to twig. Most get their start when an insect inserts an egg into a plant. In an attempt at damage control, the plant creates a bunch of cells to wall off the intruder. But the bug has a few tricks to subvert this process to suit its own needs. When it inserts its egg, it also inserts hormones and other chemical signals that prompt and shape plant growth. The insect also inserts enzymes that kill neighboring cells to create a small chamber for its larva once it hatches. The larva might secrete more of these compounds from its saliva and poop. Or, in the case of some insects that form galls in goldenrod, the larva is solely responsible for creating these growths. However it's made, the gall serves as a nice home for the immature insect until it reaches its adult form.

Not only does the gall hide the hatched larva in a cozy shelter, free from the dangers of desiccation, parasites, and predators, but the gall also acts as a stocked pantry for the insect, which eats what's around it. To help stock its gall pantry, the insect triggers nearby tissues to boost their sugar production. It also taps the plant's vascular system, drawing in nutrients from other regions of the plant. The concentration of carbohydrates and proteins in the inner layer of a gall's cells will feed the developing larva.

The gall's outer layer has a different makeup designed to deter uninvited guests, like other insects, fungi, and bacteria, and larger predators, like birds and rodents. This layer is hardened with thick-walled cells strengthened by indigestible lignin fibers and imbued with tannin, which acts as a natural

• The willow pinecone gall made by a tiny midge fly bears no resemblance to the willow tree it forms on.

antibiotic. Gall-making insects may also secrete a sugary substance that oozes out of the gall to feed ants, who then fiercely protect the gall from other predators and parasites.

Typically, galls have only a single chamber for an individual larva, but some galls are like apartment complexes for singles—multichambered, with each room housing its own juvenile bug. Each gall-making insect species has its signature architecture, making it easy to identify who made a gall and might still be hiding inside. Galls can be small blisters; potatolike structures; swellings at the ends of stalks; globe-, saucer-, button-, stocking hat–, bullet-, or tube-shaped growths; or more open, irregular bunches of leaves, flowers, or scales. Their exteriors can be smooth, spiny, furry, woolly, scaly, woody, or hairy and come in yellow, purple, scarlet, beige, chartreuse, or other colors.

A gall often differs radically in appearance from the plant tissues from which it arose. We have, for example, the willow pinecone gall, which resembles a pine cone but grows on the tips of willow branches. Some insects make mossy-looking galls on rose plants that look nothing like any part of a rose and more like something you'd find on the forest floor. (Not the same kind of gall I've spotted on the bayside rose bush this morning.)

Galls are not impenetrable. Some organisms still manage to breach them and feast on the food they provide. These uninvited dinner guests include beetles, moths, midges, and other insects especially likely to meander into some of the more openly constructed galls made from overlapping scales and leaves. One researcher examining a pine cone willow gall found close to three dozen different insects in addition to the original gall maker inside it—a mini ecosystem! These bugs may not harm the gall-creating insect but merely share its bounty, unlike wasps that have long and strong enough ovipositors (egg-laying apparatus) to inject their eggs in completely closed galls or even in the larva itself. The wasp grubs that hatch from these eggs will then kill their hosts.

But do galls usually damage their plant hosts? Some galls interfere with the production of leaves, fruits, and seeds, creating crop losses. Certain gall-making midge species, for example, significantly damage wheat, coffee, soybean, and alfalfa crops. However, the larvae of many other gall-making midges are natural enemies of crop pests and eat up the aphids and spider mites that often plague plants grown in greenhouses. The larvae of one gall midge species are widely sold in the United States for biological control of pests.

Even when they prompt hundreds of galls on a single plant or hundreds of thousands on a single tree, most gall-making insects don't seem to do any

• The spherical gall in a goldenrod stem made by a fly, whose pupa can be seen in the cross section of the gall on the right. The pupa is parasitized by a wasp.

long-term harm to their host plants, presumably because they have learned to play well with others, creating partnerships with the hosts on which they depend without significantly damaging them. (We humans still have much to learn in this regard.)

There are more than 2,000 different kinds of galls found on plants in North America, nearly half of which reside on oaks, although roses, goldenrods, and asters also often harbor gall-making insects. Experts suspect that most other plants have galls made by insects yet to be discovered.

You've probably seen galls in the stems of the wild goldenrods that paint fields and roadsides with their saffron blossoms in late summer. These galls become more obvious on the brown dried stems of the plant once they've lost their leaves in winter. If a gall is spindle-shaped, a caterpillar made it, munching away inside the gall in spring and summer and then transforming into a moth in August or September. Imagine being entombed inside a plant stem with no wiggle room for months, then squeezing through a tunnel that exits into a much bigger world. What freedom! What exhilaration! What joy, especially considering what's next—sex!

And who does the moth have to thank for its escape? Its younger caterpillar self. Before becoming a moth that lacks chewing mouthparts, the caterpillar gnawed out that tunnel to the exterior of the gall, plugging the exit hole with plant material and silk that the emerging moth can easily push out. Without the caterpillar's diligent preparation, the moth would have been trapped inside the gall.

After leaving the gall and mating, a female moth lays her eggs in autumn on the lower leaves and stems of goldenrod. Hatching from the eggs in spring, the larvae crawl around until they reach new goldenrod shoot buds, which they burrow through and beyond to create galls in the stem.

If a gall you spot on a goldenrod stem is spherical, it was likely made by a fly whose larva, like a moth caterpillar, has chewing mouthparts that its adult version lacks. In the fall, before going dormant for winter, the grub chews out a tunnel stopping just short of the outermost layer of the gall, which stays intact, creating a weak window. It then retreats to wait out the winter in the gall's center.

If the larva survives until spring, it becomes an adult fly equipped with a cool airbaglike structure on its forehead. With dramatic flair, the fly pumps its bodily fluids to make that sac rapidly expand and punch a hole through the gall's window. *Bam!* Freedom! The remarkable punching balloon used to enable its escape soon recedes back into the insect's head, never to be expanded again. The tawny-colored adult gall fly sports speckled wings and is a little smaller than a housefly. Sticking around its goldenrod home, the fly has only two weeks to live, during which its sole purpose is to mate and, if female, lay eggs in goldenrod plants with her piercing ovipositor.

There are several different species of goldenrod, and this female gall fly is picky. Once she has mated, she must find her own particular goldenrod species on which to lay her eggs. How does she do that? With chemical sensors on her feet and antennae that can detect minute differences between species. Even when researchers tried to fool these flies by wrapping goldenrod leaves from one species around the stems of another, the flies still laid their eggs on their own special species. Ten days later, cream-colored grubs emerge from the eggs, triggering the construction of golf ball–sized galls that take nearly a month to complete.

During our field study, Charley showed my class another type of goldenrod gall I wouldn't have noticed otherwise because it's so miniscule. This yellowish-green elongated swelling was on the end of a threadlike stalk extending from the flowering portion of the plant. I popped this gall open and was shocked to see a tiny bright neon-orange grub inside. Charley told us that it would mature into a flying midge no bigger than a fruit fly. Recent DNA studies have revealed a vast number of these previously overlooked miniature flies, suggesting that they may include as many as one million species yet to be described, which would make them the largest family in the animal kingdom. Isn't that poetic justice, having one of the tiniest flies comprise one of the biggest families? But what good is a large family if it can't protect you from all the many predators that want to eat you? Hence these bitty grubs hide out in galls in goldenrod flowers and have become the insect equivalent of flower fairies.

Unlike the miniscule midge goldenrod galls, oak apple galls are about the same shape and color as an apple when the galls first form in spring. On a field trip with Charley, I plucked one of these galls from an oak leaf.

• An oak apple gall; the opening reveals the inner larval component held in place by filaments to prevent wasps from parasitizing it.

Because it was summer, it had already transitioned to a boring beige color, mottled with brown. I cracked it open and found hidden inside a round seedlike structure held in the center of the gall by a dense mat of long, thin fibers. It was one of the most bizarre things I'd ever seen in nature.

"What's going on here?" I asked Charley, and he explained the gall wasp that makes this type of oak apple gall protects its developing larva from other wasp predators by cleverly suspending it in a small inner casing in the center—the seedlike structure—and tying it to the outer gall covering with multiple radiating filaments. Any parasitic wasp that comes along afterward is not likely to have an ovipositor long and strong enough to breach the inner larval compartment and lay its own egg inside it or the larva.

Oak trees are a favorite for so many types of gall-producing insects that a single tree can host dozens of species, which in turn feed other wildlife. For the most part, no oaks are harmed in the making of these galls. Who knew so many insects inhabit these trees?

But back to the rosebush by my feet. It turns out that a wasp made these galls, and their spines protect the wasp grubs from predators, especially munching rodents and birds. If they survive the winter, their adult forms will chew their way out of the galls in spring or summer. (Unlike moths and flies, wasps have chewing mouthparts, so they don't need their grubs to prepare escape tunnels for them.) These fruit fly–sized wasps have distinctive humpbacks seen only under a magnifying lens, and they live less than a week or two. During their brief lives, the wasps mate and females lay their eggs in emerging leaves or in the rose plant's swollen buds, with gall formation starting before the eggs hatch. So you're more likely to spot the gall than the miniscule wasp that made it.

I wish the gall wasp larvae well, hoping they don't get swept away by some winter storm or eaten by wasp parasites or other predators. And I thank both them and Charley for introducing me to a whole other world I hadn't realized existed, a world whose inhabitants spend most of their lives enclosed in the stems, leaves, and other parts of plants due to their remarkable ability to hijack a plant's growth instructions. It's a working partnership between insect and plant, serving as yet another marvel of nature. That's what's so cool about insects—they're the end result of so many millennia of evolution intertwining completely different organisms and crafting incredibly intricate adaptions.

LISTENING TO INSECT SERENADERS, PART 1

Katydids (Family Tettigoniidae)
LOCATION: On a *Salvia* plant in a garden, northeast Maine
DATE: September 1 TIME: 3:49 p.m.
(fork-tailed bush katydid, *Scutteria furcata*)

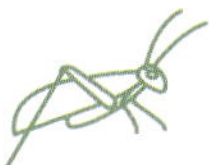

It catches me by surprise—a spring-green leaf resting on a blue salvia flower in my Maine garden in early September. I look closer and notice that this "leaf" has two long antennae poking out and a pair of oversized bent legs creating triangle shapes on each flank. This is no leaf—it's a katydid!

I recognize it because, years ago, I had the unusual good fortune to spot one of these well-camouflaged insects hanging out at night in our suburban yard in Pennsylvania, where my husband and I lived before moving to

• My daytime sighting of a fork-tailed bush katydid surprised me.

Philadelphia. Whenever we sat on our side porch during warm, late summer evenings, we heard their loud and distinctive *katy-did* chants.

But except for that one time, we only ever *heard* the katydids; we never saw them. So the one I've spotted on the salvia shocks me. I thought these insects were active only at night and usually heard but not seen. This katydid is the opposite—seen but not heard, and active during the day. Our Maine evenings lack the katydid's namesake chatter; why? The insect serenaders up here also seem to lack the more musical chorusing of crickets, instead emitting a continuous high-pitched trilling bordering on buzzing. Who's doing this remarkably persistent chanting night and day, how are they doing it so loudly, and where are they? How can something ever present to the ears be ever absent to the eyes?

All these questions trigger a venture into the incredibly diverse realm of insect serenaders. It's astonishing how these never-tiring sirens of forest, meadow, and marsh make and amplify their sounds, court mates, and usually keep out of sight while tantalizing us with their chants and synchronized choruses.

First I'll tell you what I discovered about that katydid I saw in my Maine garden.

It's called a fork-tailed bush katydid, but we'll call it F.K. for short. You usually find it and other bush katydids in shrubs or other low-lying vegetation bordering woodlands, unlike the more raucous common true katydid species, which we'll call C.K., that is seen farther south. C.K.s hang out in the tree canopy.

By the end of summer, our F.K. has grown his sexual parts and large leaflike wings so he, akin to Walt Whitman's poem, can finally sing his song of himself, his mating call. And wow, what a Romeo he has to be to get it on! Unlike the courting seen in most insects, with bush katydids, the females take the initiative, and they require a good concert, plentiful dining, and long foreplay for males to have their big night.

After sidling up to a male whose call attracts her, if he passes muster, the female katydid mounts him from behind. In many species, once she's mounted, the male uses rapid and rhythmic tapping for as long as half an hour to prompt her genital area to open. Once it does, there's access to the sac where she stores sperm to fertilize her eggs. He then ejaculates his packet of sperm and attaches it to the tip of the female's abdomen.

"There's stiff competition between guy katydids to be the first to get the gal, who relies heavily on a male's song to judge his fitness as a mate."

Next comes the dinner date. The male katydid covers his sperm packet with a frothy nutritious substance the female eagerly consumes. This dinner keeps the female busy long enough for the male's sperm to travel down a tube connecting his sperm packet to her sperm storage sac. The larger the meal, the more progeny she can produce. Since bigger males tend to provide bigger meals, females prefer to mate with them. And when there's scarce vegetation to dine on, females mate with more than one male. Males down the line having sex with a female aren't likely to fertilize her eggs, though, because in katydids, it's first come, first served, fertilization-wise.

Consequently, there's stiff competition between guy katydids to be the first to get the gal, who relies heavily on a male's song to judge his fitness as a mate. The louder and longer, the better, as those qualities often equate with size. The female hears these love songs with ears located on her knees, which extend far above the vegetation she rests on because of her long, bent legs.

Like teenage girls, female katydids tend to be drawn to the long-winded lead singer in a band when a bunch of katydids sing. This is the one who makes the first call, and other nearby katydids follow him with their own songs. Male katydids often band together while calling females, giving them safety in numbers by providing a curtain of sound impenetrable to predators seeking an individual to eat. A large male choir also draws more females. But chorusing in both katydids and crickets mainly happens due to competition between males to be that sexy lead singer. "The resulting synchronicity . . . evolved as the result of an individual's attempt to call attention to itself," noted John Himmelman in his book *Cricket Radio*. He mentioned a study of crickets that found that, "with all the males attempting to be the lead-calling cricket, the male chirps were just milliseconds apart, which to our ears sounds as if they are blended as one." The end result is a throbbing rhythm, nature's late-summer heartbeat auditorily conveying the sexual yearnings of hundreds to thousands of these insects, each hoping to get a chance at having sex and continuing their lineage during the few weeks to months for which they're winged adults before dying.

A katydid will even competitively chirp with a typewriter.

While punching away on an old manual one (remember those?), entomologist Richard Alexander noticed that when he pounded out a staccato triplet rhythm similar to that of a C.K. caged nearby, the katydid responded with his own triplet. When Alexander slowed or sped up his typing tempo, the katydid did too. This katydid viewed the typewriter as his competitor and made sure he was heard in between its tapping. When Alexander added an extra tap to his typewriter rhythm, the katydid extended his triplet call *ka-ty-did* to a quadruple one—*ka-ty-did-did*. Alexander continued to play with this katydid's call, eventually getting it to add as many as three extra *did*s. A study in another species of katydid found that a male

• I spotted this katydid at night dining on a *Rudbeckia* flower in my garden.

starts calling out with just one or two chirps, but will, as it progresses, keep adding a chirp to its call (without any prompting of a typewriter) until it has finally reached its capacity and pauses. The male that achieves the highest number of syllables—the lead singer trying to outchirp his competitors—has the greatest chance of being chosen by a female. Do you realize what this means? Katydids can count! (Although not to more than four, some studies suggest, so brilliant mathematicians they're not.)

The bush katydid in my garden (F.K.) wasn't making a sound when I saw him. Perhaps it was too early in the day to start serenading the ladies, or perhaps he was intimidated by me peering over his shoulder? All katydids and crickets make their music by rubbing the toothed veinlike structure on the underside of one wing against the raised smooth ridge of the other wing, like dragging a thumbnail over the teeth of a comb. A drum-shaped structure next to this ridge loudens the scraping sound.

But each type of katydid or cricket has his own unique song. By listening to recordings I found online, I discovered that F.K.'s song sounds *nothing* like C.K's. C.K. makes frantic rapid trills by opening and closing his wings rapidly in quick succession, creating a regular percussive rhythm. In contrast, F.K.'s sound is more languid and soft-spoken, the cool cat playing bass in the band, eliciting a single *tsip* every ten to fifteen seconds, sometimes bunching up several *tsip*s between long pauses. F.K. also tends to give a series of (excited?) high-pitched ticks when encountering others of his ilk.

For the rest of the summer, I am on high alert for that subdued call and one day finally hear those intermittent *tsip*s coming from the edge of the forest in my yard. Eureka! I can't actually *see* the bugger who is sounding off, of course, as he is so well camouflaged and deft at hiding, but his call is clear. Why, I wonder, have I never heard it before?

Because I wasn't paying attention.

"We Americans are too busy, too hurried, too mechanized in our moods, to hear what this cricket or that katydid says. . . . We are a hurried, worried people," mused naturalist H. A. Allard in the July 1928 issue of *Scientific Monthly*. And when we do notice things, it's usually with our eyes rather than with our ears. We have natural "aural dampers," which tune in to what's most salient and block the rest. "Because of this, we miss out on a most extraordinary concert being performed from the trees, shrubs, and fields

• Each katydid species has its own unique song. This short-winged meadow katydid makes a series of faint, high-pitched ticks followed by longer buzzing.

that surround us. . . . That is a gift we humans have been given. . . . [These] sounds can enrich our lives," wrote Himmelman.

The Chinese and Japanese seem to know this, having a centuries-long tradition of keeping crickets in cages to enjoy their songs. In his book *Bug Music*, David Rothenberg recounts a story told by an acquaintance who, while standing in line at a Beijing cricket shop, was listening to recordings of American crickets. He passed his headphones to the Chinese man behind him, whose eyes lit up upon hearing the cricket concert. When told that most people in the United States don't appreciate cricket songs, the man became upset: "Those stupid Americans, they have even better crickets than we do, and they don't appreciate them?!"

Well, I'm determined to use both my eyes and ears to appreciate all these crickets and katydids and to better understand the insect choir providing the soundtrack to my late summer and early fall months.

LISTENING TO INSECT SERENADERS, PART 2

Fall Field (*Gryllus pennsylvanicus*) and Other Crickets

LOCATION: Under a log in a forest, northeast Maine

DATE: September 20 TIME: 3:39 p.m.

It's the end of September in Maine, and I don't hear the birds' love songs anymore. Even the sounds of the winter wren, whose lovely bubbling waterfall of notes was ever present while I gardened earlier in the season, now rarely fill the air. Instead, insects are having their moment. The air is filled with a continual high-pitched ringing sound that becomes louder at night as other sounds die down for the day. Accompanying this pulsating chorus is a sense that something momentous is about to happen, like the drumroll before a big announcement—*And the winner is. . .* Is it

• Fall field crickets, such as this one, are usually heard but not seen because they are so elusive.

the drumbeat presaging winter and the sense of mortality it brings? The throbbing heartbeat of life itself that you become more aware of just before it's extinguished?

Something major *is* about to happen. Nature's summer bounty is about to leave us. The maple leaves will yellow, the light in our yard will lessen, the birds will head to warmer climes—and the world will end for many insects in temperate zones.

These serenading insects are singing summer's requiem.

And what are they telling us—is there a deeper message we should heed?

I follow some chirps to catch one singing. As soon as I close in on the call, it stops. There's so much sound in the air, I don't know where to stalk next. They are everywhere and nowhere at the same time. "The singing insects are wary, alert, clever in the art of concealment, and as deceptive as the most skillful ventriloquists," noted Vincent Dethier in his book *Crickets and Katydids, Concerts and Solos*. Dethier recommended that human stalkers be stealthy and patient and understand the singer's idiosyncrasies so we know where and at what time to best find him.

I take his advice and read that fall field crickets are common, large, and loud. These crickets don't start singing until mid-July to early August, when they've finally grown wings and develop an adult sex drive, prompting them to chirp their presence to potential mates. They do this both day and night until they're killed off by the cold. Fall field crickets hide under logs, stones, or dead leaves. I hear some chirping nearby while walking on the edge of the forest that abuts our yard. As soon as I stop and crouch down to get closer, the singing stops, of course. But when I turn over a large piece of bark on the ground, five big crickets, like shiny black bullets with legs, scurry away to other hiding spots, leaving behind one who froze, hoping to escape my notice. I snap some photos. He doesn't chirp while I'm there, so I don't get to see his wing scraping action, but I solve the identity of at least one of my yard's serenaders.

After enlarging his photo, I notice a pair of unusually long tapered spines sticking out of the cricket's rear end. These spines, which are more than half as long as his body, detect nearby movement from alterations in air currents. No wonder silent stalking doesn't work with these guys. Even if you're careful not to crunch on any leaves or make other noise while approaching a cricket, these spines, called cerci, will detect your movement at a distance. The cerci are covered with hundreds of extremely sensitive hairs—a hundred times more sensitive than our eyes' light detectors. These incredibly good motion detectors explain why these crickets went mum as soon as I closed in on them!

Sadly, these crickets were the only ones I was able to detect, despite there being more than two dozen cricket species in Maine. They can be black, like the field cricket, or pale green or brown, and they come in all shapes and sizes. Some, like katydids, resemble leaves, while others mimic stems or buds in their coloration and shape. Some chirp in unison at regular intervals, while others trill continuously and sound more buzzy. Some live in sand or in muddy burrows, while others live in bushes, grasses, or trees.

My favorite is the snowy tree cricket, perhaps the most familiar—this movie star's chirps often appear on film and in songs. Unlike katydids, which sing a wide range of pitches simultaneously, giving their songs a buzzy quality, the snowy tree cricket has more focused pitches that sound melodic, with a comforting regular chirping rhythm. Snowy tree crickets are also usually the loudest of the insect chorus because they outnumber the other serenaders and they sing from trees, so their songs easily drift down into our ears.

What makes snowy tree crickets especially endearing is their ethereal appearance. These forest fairies have greenish-yellow bodies with diaphanous,

• The snowy tree cricket's chirping has been featured in several popular songs and movies, as it provides the soundtrack of late summer.

rounded, flat wings that take on a heart shape to amplify their sound when they serenade. To raise their volume, they chew a hole through the middle of a leaf and poke their head and wings through it—the leaf becomes a soundboard. A mole cricket goes so far as to dig his burrow in the shape of a megaphone and then sings from the opening.

Clever crickets.

I'd love to see a snowy tree cricket, but despite being so common, they're hard to spot as they're usually perched high in trees, where they're well camouflaged. They also hang out in deciduous forests, not in the evergreen forests we have at our place in Maine. Here I'm more likely to see the pine tree cricket, which hides in pines, spruces, and other coniferous trees. After listening to its song, I suspect the pine tree cricket is making the persistent ringing I've been hearing at night. This small and slender cricket has a russet-colored head and body topped with green wings so similar in coloring and shape to its surroundings that it's hard to detect. Plus, the calls of pine tree crickets bounce off surrounding branches, making them hard to locate, especially in the dark. "We must therefore rely upon our ears," advised John Himmelman, author of *Cricket Radio*. "To truly experience what these insects can contribute to our aesthetic pleasure, we must listen."

I'm disappointed, but there's something to be said for leaving a bit of mystery in the air. In *The Sense of Wonder*, Rachel Carson wrote of not being able to identify an insect serenader she and her nephew heard one summer night in Maine: "I'm not sure I want to. His voice—and surely he himself—are so ethereal, so delicate, so otherworldly, that he should remain invisible, as he has through all the nights I have searched for him."

Viva le mystère. Simply knowing that I'm outnumbered by hundreds to thousands of different soloists and choristers is enough. As Dethier asserted, "We human beings are not necessarily the most important inhabitants of the planet . . . to hear a katydid is to have a more balanced perspective on our place in the cosmos."

Both crickets and katydids tend to live only until the first frost. That's why we link their songs to summer's end, as we do the wild goldenrod and purple asters that dab the landscape with color, and the saffron, crimson, and russet leaves that float to the ground. The sounds mark time passing, adding auditory signals to visual ones that tell us winter will soon be among us. As Himmelman noted, reflecting on the trilling of a lone ground cricket in November after most insects had died, "There are no females left to answer. It doesn't matter. It is trilling because it has to, and it is giving it everything it's got. It is the violinist playing as the *Titanic* is sinking. It's what they do. How can that not stir a soul?"

I think I've deciphered the message these insect serenaders are sending us: *Pay attention. We are here all around you. Many of us.*

Until we're not.

REFLECTING ON INSECTS IN WINTER

Green-Striped Grasshoppers (*Chortophaga viridifasciata*) and Many Other Insects

LOCATION: Garden in northeast Maine

DATE: November 10 TIME: 10:58 a.m.

It's November in Maine. Although frost hasn't yet killed my garden perennials, most have died back or sport fall colors. Maples and birches have lost their leafy garb, and the low-angled morning light imparts a luminous glow to the ambered sedges and marsh grasses edging the bay. Not surprisingly, I see and hear few insects. Even the cold-tolerant bumble bees aren't buzzing about with no flowers to ply.

I suspect both near-freezing temperatures and lack of food has forced most insects to disappear from our area, something the migrating birds started doing months ago. I miss seeing the insects' brilliant flashes of color

• Unlike many insects, this two-striped grasshopper will overwinter just as it is, instead of as an egg.

and hearing their soothing drone. Instead, I hear only the chickadees' frenetic chattering and the woodpeckers' pounding. What do these year-round Maine residents eat once insects vanish and seeds sink into the ground?

I always find the world outdoors in fall and winter rather lonely, with few sightings of animals except for deer, who relentlessly return to the garden for any scrap of food remaining. But I also appreciate the pause in the usual quick pace of life that cold weather gives us. Other seasons overload me with sensory stimuli and the work that must follow: After seeing an iridescent beetle flying, I must photograph it; after hearing a bird warble, I have to find it; and when I spot tiny pink eggs the size of rice grains glued to a periwinkle shell, I'm compelled to learn who laid them. Sights and sounds besiege me while hiking, biking, or kayaking during warmer months. Sometimes all those stimuli exhaust me more than the physical exercise.

Once winter arrives, I'm happy to have time to reflect on all I've encountered in other seasons. Spring, summer, and fall provide an action-induced blur of a verdant landscape, as if viewed through a moving car window. Winter, in contrast, offers a crisp and clear black-and-white frame—a simpler picture to probe. It helps me discern what to highlight and what to crop, what to keep and what to drop. During winter I turn inward, crystalizing the essence of what I've experienced.

While doing the final weeding before tucking flowerbeds under a winter blanket of leaves and seaweed, I notice something tiny hopping away from me. It takes me a while to then find this brown grasshopper, as it's well camouflaged against brown leaves and soil. But eventually I see his large round eyes peering intently at mine and take a picture. (I suspect it's a he because even though it's called a green-striped grasshopper, *Chortophaga viridifasciata*, only the females tend to be green.)

This grasshopper isn't flying away with the typical clattering sound. Instead he stares at me, holding a frozen pose. I welcome his presence, recognizing how rare he must be at this time of year. Later, looking into his species, I learn that he's a *juvenile* grasshopper, and that he will overwinter just as he is in Maine, where temperatures often dip below zero.

"I discover that grasshoppers and many other insects have superpowers that enable them to withstand freezing."

How's that possible?

I discover that grasshoppers and many other insects have superpowers that enable them to withstand freezing. An Arctic woolly bear caterpillar, related to those brown-and-black furry caterpillars often found crawling farther south in fall, stays frozen for all but two months of the year and can survive temperatures lower than -70°F. And the fly larvae that cause those spherical gall swellings in goldenrod stems survive even *lower* temperatures. Even lacewings, as delicate as their name implies and no bigger than half an inch long, can withstand temperatures as low as -40°F in the Arctic. How?

I take Charley Eiseman's online "Bugs in Winter" class to find out.

In the class, Charley explains that though we don't usually see insects flying around in northern regions of the country in this frosty season, most don't abandon their habitats once the weather turns chilly. Instead, they hunker down and wait out winter as eggs, larvae, juveniles, pupae, or adults, usually in a protected spot. To survive when food is scarce, many

• Delicate lacewings, like this one, are no bigger than half an inch long, yet they can withstand temperatures as low as –40°F.

insects experience a type of dormancy called *diapause*, which suppresses their metabolism and development. Well-known examples are butterfly and moth caterpillars that, prior to their winter stillness, construct cocoons or chrysalises in which they rest as pupae during winter, often buried under snow, leaves, or some other form of insulation.

Other insects, like certain beetle larvae, avoid the freezing cold during their diapause by burrowing deep underground below the frost line. Aquatic insects like caddisflies and mayflies move to deeper water that doesn't freeze, where they continue to feed and grow in winter. But the rest of the insect gang diapausing in bark crevices, under leaf litter, or hidden in some other sheltered spots still need a way to protect themselves from freezing to death.

Remember the White Witch of the Narnia series, who quickly spreads ice over everything she touches, or more recently Elsa of the Disney movie *Frozen*, who does the same? Something similar happens in the real world. When temperatures dip below freezing, dust and other particles seed ice formation, acting as ice crystal "nuclei" upon whose surfaces water vapor condenses and then freezes. This ice spreads rapidly as soon as it starts, the lacy crystalline network extending outward and freezing everything it comes in contact with.

The ice is deadly. Inside cells, it pierces and compresses essential structures. But insects have several strategies to elude harm induced by freezing temperatures. One way is to boost their production of sugary compounds, nature's antifreeze. The more they build up these compounds inside their cells, the less likely water in their bodies will freeze. The compounds also prevent harm by substituting for water molecules in the insect's enzymes and cell membranes so they continue to function properly.

Other insects avoid freezing by avoiding contact with ice. Yellow jacket wasps in Alaska do this by dangling from the undersides of leaves or leaf litter, latching on with their jaws. Imagine spending your winter like an animal hung from a meat hook in the deep freeze at a butcher's shop. Another insect freeze-avoidance strategy is burrowing into dry spots that are free of the dust and other particles that can seed ice formation.

As for that green-striped grasshopper, it can be active even into February or early March in northern climes, surviving by feeding on evergreen grasses, such as sedges. But being active in winter is challenging. So for at least part of the season, until days lengthen, these grasshoppers go dormant under dead vegetation in meadows and lawns. They too need a temporary stop-action on life, a rest from the frenetic search for food.

While diapausing, green-striped grasshoppers can survive temperatures below zero thanks to their antifreeze compounds. When spring arrives, they shed their exoskeletons for the last time, becoming adults and gaining wings. Because they are the earliest grasshoppers on the scene in spring, they pig out on all the grassy food available to them until their cousin species, hatching from overwintering eggs, grow big enough to have a seat at the table. Is it worth the long-frozen wait to have this head start on dining? Only the grasshopper can tell you, although there must be some advantage to this strategy or it wouldn't have evolved.

So even in winter, insects surround us—they're just hard to find. But wintering birds know where to look. Woodpeckers snack on diapausing ants and peck out fly and beetle grubs from trees, deadwood, and goldenrod galls. Kinglets, walnut-sized birds I often see flitting about like moths, munch on the diapausing inchworms they find on various trees.

• Because they overwinter as juveniles, two-striped grasshoppers get a head start on feeding in spring over their cousin species that overwinter as eggs.

I don't follow these birds' leads to discover insects in winter. Instead I take advantage of this still season, soak up that restorative pause in activity that will ready me for future insect safaris once the weather warms. There will always be something new under the sun to discover—I just need to look and listen for it.

IF YOU PLANT IT, THEY WILL COME

This spring I'm gardening for grubs.

After more than thirty years of gardening for flowers to fill my vases, fighting off any insect pests hampering their growth, I've changed my tune. I'll still continue to grow perennial flowers, including native bloomers like purple coneflowers, butterfly milkweed, and bee balm, as well as showier nursery-bred varieties, like double-blooming daylilies, delphinium, and bearded iris. But starting this year, I'm trying to be more ecologically aware, both in what I plant and in how I tend my gardens and yard, to provide more habitat for insects and the birds depending on them.

Birds and insects are both in serious decline. The data on insects are not complete and vary from region to region and insect to insect. But disturbingly, one study reported in the scientific journal *Biological Conservation* found that, worldwide, about 40 percent of insect populations are declining so precipitously they could be extinct in a few decades.

• My perennial flower garden features many insect-attracting native plants, such as the orange butterfly milkweed in the foreground.

Other studies find nearly one-third of North American bumble bees and one in five US butterfly species at risk of extinction, as well as nearly a quarter of all butterfly species in the US declining in number since 2000. Multiple factors are behind these declines, including overdevelopment and agriculture, nonnative plantings, pesticides, light pollution, and climate change. The purported "insect apocalypse" has been accompanied by about a 30 percent decline in the number of songbirds spotted in temperate zones since 1970, as reported in *Science*. Most of these birds are borrowed from the tropics and migrate up here, blessing us with their flashy colors and catchy songs only if they're repaid with bountiful sources of food, especially high-protein insects for their growing nestlings.

So I'm taking advantage of the advice and online resources of the Xerces Society for Invertebrate Conservation, as well as minding findings from recent research on what plants and practices foster the most diversity and abundance of insects and birds. I'll share what I've learned, hoping you too might be inspired to do what you can in your yard, no matter its size, to boost the number and diversity of bugs and birds.

Residential areas can provide important food and shelter for many threatened and endangered pollinators. By establishing pollinator habitats in your own yard that give them a safe oasis, you will be an active part of restoring species on the brink. Even what little you can plant in a city

• A bumble bee dines on nonnative globe thistle growing in my garden.

courtyard, balcony, or window box can be a lifesaver for some species. Recent studies suggest that pollinators may actually do better in urban and suburban environments, where they're less exposed to agricultural cultivation practices that destroy their food sources and saturate the environment with pesticides deadly to insects.

While trying to do my part in providing for insects, I'm keeping in mind three main components of a good insect habitat:

- Plentiful diverse food resources during the entire growing season
- Sheltered areas in which insects can nest, rest, hide from predators, and find protection from winter cold and ice
- Protection from pesticides, including herbicides

Let's tackle food first. Scientists currently debate whether insects will be more abundant and diverse in a garden planted solely with species native to the area versus a garden with native species as well as closely related nonnative plants or more exotic ornamentals. Most insects dine on plants, and experts theorize that because insects evolved in tandem with the plants native to their area, the insects are well suited for tapping these plants' nectar and pollen or dealing with toxins in their leaves. This is especially true for most butterfly and moth caterpillars and more than a quarter of bee species. It's also indirectly true for many predatory wasps, which tend to specialize on certain types of insect prey that in turn will dine only on certain native plants. But many insect species (about 10 percent) are generalists capable of accessing food from multiple types of plants.

And it's not just garden plants insects dine on. Trees and shrubs comprise two-thirds of the host plants for butterflies and moths. Several studies have found that red maple, black cherry, elderberry, spicebush, viburnum, and other native trees and shrubs support many more moth and butterfly caterpillars than commonly planted nonnatives such as ginkgo, crepe myrtle, Bradford pear, and forsythia. Oak trees are a particularly good nursery for an astounding number of moth and butterfly caterpillars nationwide—more than 950 species. In plots with lots of different native trees and shrubs, insects are not only more diverse but more abundant, and the same is true for birds.

So, after consulting lists of plants native to my area and beneficial to insects and birds provided by the National Wildlife Federation and the Xerces Society, I decided to plant native elderberry, viburnum, and clethra shrubs, as well as more native flowering perennials, such as New York ironweed and swamp milkweed, in one of my recently expanded garden plots. These plants will add to the beauty of my flower arrangements while supporting more specialized insects.

But I'm not going all native in the garden, because many insects can tap food resources of plants closely related to native species. Some studies find that insects, especially bees, are just as abundant on these nonnative plants as they are on native species, and the more important factors drawing flying insects to gardens are the amount of sunlight and total floral area. In some cases, even the more exotic flowering species plant nurseries frequently sell can help these pollinators by extending the growing season earlier in spring or later in fall, when most native plants aren't blooming. Also, by flowering abundantly throughout the growing season rather than just for the few weeks to months for which most native plants are in bloom, some of these nursery plants can give an extra nutritional boost to insects that are generalists.

I've noticed several different species of native bees plying the flowers of the pot of petunialike calibrachoa flowers I keep on my deck, and I was especially surprised to see a sphinx moth tapping its nectar at night after visiting its preferred long-tubed tobacco flowers growing in an adjacent pot.

• A bumble bee drinks nectar from verbena flowers in my garden. This plant, *Verbena hybrida*, is native to South America, but still attracts bees, butterflies, and other insects.

Calibrachoa flowers are native to South America, yet that doesn't stop some insects in Maine from enjoying their exotic fare.

But not all flowering plants sold in nurseries will attract insects. I have noticed that the native fall-blooming asters are a lifesaver for flower flies, bees, and wasps that are still active in autumn, whereas fall mums don't seem to draw these pollinators. Unfortunately for the pollinators in my vicinity, deer always seem to eat my aster blooms, while leaving mums alone. So I've seen several bee species resort to visiting my pot of calibrachoa flowers instead during the fall season. As Lars Chittka noted in *The Mind of a Bee*, many so-called specialized bees will exploit other flower species when their typical food becomes scarce. "Upon close inspection the purported affinities for certain foods are often less strong or exclusive than was once supposed," he wrote.

In addition to planting more insect-drawing plants, I also stopped automatically pulling weeds from my gardens, recognizing that many of these plants provide food for insects. Not only do I no longer weed out violets, having learned that they are the sole food for fritillary butterfly caterpillars (see "Finding Tiny Miracles Hidden in Plain Sight," page 37), but I closely observe the weeds in my garden, assessing whether they have a lot of insect visitors. If they do, I keep them in my plot unless they encroach on the territory of my more favored bloomers. Yes, it looks messier with all these weeds in between my perennials, but I recognize that nature is disorderly too, and I'm willing to sacrifice a neat appearance if it brings more insects and birds into my yard.

"I'm trying to be more ecologically aware, both in what I plant and in how I tend my gardens and yard, to provide more habitat for insects and the birds depending on them."

Moving on to shelter, insects need places to hide from their predators, especially while they are growing into their adult forms or waiting out winter. Some, like many native bees, flies, beetles, and wasps, nest and/or develop underground, while others, including luna moths and swallowtail butterflies, carry out their larval or pupal lives wrapped or embedded inside leaves or stems or attached to other vegetation. Others hide in deadwood or in tree cavities.

Cognizant of these insect proclivities, I no longer bag and dispose of fallen tree leaves in my yard. Instead, I rake them into my garden plots, where they give added insulation over the winter to both dormant insects and garden plants, while also helping to suppress weeds and retain moisture. Such leaf mulch is as good as wood mulch and provides a better alternative because a thick layer of wood mulch hampers ground-nesting insects. I never shred leaves with a mower to avoid shredding insects hiding inside them.

I also no longer remove my garden plants' dead stems and seed heads in fall, as they can feed birds and house overwintering insects. Instead, I wait until late spring to cut them down. I do so at varying lengths starting at eight inches above the ground, rather than pulling them out altogether, in case insects are still inside them, and to offer new nesting sites for newly emerging bees and other bugs. (Insects vary as to where inside stems they find shelter—hence the different lengths.) New vegetation quickly grows above last year's stems, making them inconspicuous, and I enjoy seeing if any of the stems become plugged with leaves, mud, or pebbles, suggesting insects hiding inside. Although there are commercially available "bee hotels" made from bamboo or blocks of wood, the Xerces Society explains that these may lead to a buildup of diseases or mites, especially if they're not thrown out and replaced after a season or two.

Snags and stumps are important "hotels" for numerous insects, including those beautiful longhorn and jewel beetles, as well as cavity-nesting bees. So I leave this living deadwood on my property for the bugs and the birds that feed on and nest in it. I also take pleasure in knowing my piles of firewood

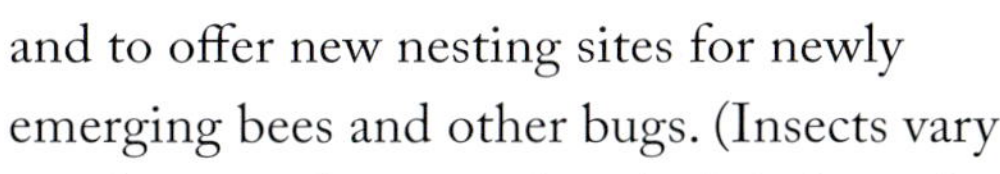

• Bees and a monarch caterpillar dine on butterfly milkweed in my garden. Many insects, like the monarch, will only feed on specific native plants during their larval stage.

and garden brush offer firefly larvae shelter during the day and throughout winter, and might provide an insulated home for mourning cloak, woolly bear, and great spangled fritillary butterfly caterpillars, as well as other insects overwintering in our area.

Finally, I try to give insects protection from the harmful effects of pesticides prevalent both in rural agricultural areas and in local urban and suburban parks. Even when they don't kill insects outright, research reveals that pesticides decrease the reproduction and navigation abilities of pollinators, as well as weaken them over time, making them more susceptible to disease. Studies find that the common herbicide 2,4-D, often included in weed-and-feed products, kills or weakens a wide variety of insects, including honey bees. One disturbing study found that only 6 percent of spicebush swallowtail eggs hatched when exposed to the common and long-lingering herbicide glyphosate (Roundup), compared to 100 percent of the eggs that hatched from leaves not exposed to it. And keep in mind that weed killers kill not just weeds but the insects depending on them for food.

So I don't use any pesticides in my garden and yard, instead removing unwanted weeds and insects by hand. I'm also changing my attitude about insect damage in the garden. Yes, leafhoppers yellow the plant leaves they suck, but that doesn't stop the plant from blooming, and I love seeing these cute little bugs' neon-green and carmine stripes. I've realized that the holes caterpillars and other munchers make in the leaves of my flowering plants usually don't significantly harm them. Even when monarch caterpillars and tussock milkweed moths stripped my butterfly milkweed plants of nearly all their leaves, the plants regrew them, or they grew back the following year. And I remind myself that any damage done to the beauty of my flowers is compensated by the beauty of the insects and the melodic songbirds they'll feed. I'm especially willing to sacrifice the leaves on my butterfly milkweed

• A purple coneflower nectar bar with two customers—a skipper butterfly and a tricolored bumble bee.

plants if that leads to more monarch butterflies, which are increasingly rare these days.

For so long I saw insects munching on my flowers as the bane of my garden's existence. But now I purposely plant for their presence, recognizing that I can grow both gorgeous flowers *and* insects. *If you plant it, they will come* is my new motto. I hope that with my new plantings, the bug bounty in my garden will expand, bringing with it more warblers and other brilliantly colored songbirds from the tropics, providing enchanting beauty and background music while I garden.

I can't imagine the growing season without them.

Glossary

aeolian zone—the portion of Earth's atmosphere, nearly as high as the clouds, where strong insect-carrying winds predominate.

caterpillar—the larva of a moth, butterfly, or sawfly.

cerci—the motion-detecting pair of spines projecting from the rear end (abdomen) of an insect.

chrysalis—the hardened cover a caterpillar forms to encase itself while it transforms (pupates) into a butterfly.

cocoon—the silk encasing the larval form of an insect while it pupates into its adult form.

complete metamorphosis—the type of life cycle followed by beetles, butterflies, flies, and other insects in which eggs hatch into larvae, which develop into pupae, which eventually become adults, typically with wings and reproductive organs.

diapause—a temporary pause in metabolism and development to avoid adverse environmental conditions such as winter cold.

• A sweat bee in a *Rudebeckia* flower.

elytra—the outer hardened pair of wings of a beetle that encase its other pair of wings, which is used for flying.

exoskeleton—the external covering (skin) of an insect that gives it support and protection. Insects' exoskeletons are typically strengthened with tough, hard fibers known as chitin.

family—a midlevel taxonomic grouping of insects with similar characteristics that includes several genera.

gall—abnormal growths on plants in response to infiltration by insects, mites, or microbes.

genus—a low-level taxonomic grouping of several insects with similar characteristics. An insect's scientific name is comprised of its genus (which is capitalized) followed by its species' name (lowercase). Both are italicized.

grub—the immature (larval) form of many beetles, although the larval forms of bees, wasps, and flies are also sometimes called *grubs*.

heartwood—the dark inner core of a tree that is comprised of dead fibrous cells that provide support.

hilltop mating—the gathering of insects at hilltops in order to find mates.

honeydew—the sugary liquid that aphids, spittle bugs, and other insects excrete while feeding on plants. Honeydew is a source of food for ants, who may tend the insects that produce it.

imaginal discs—the "budding" cells of the larva that will develop into key structures, such as wings, legs, or antennae, in the adult insect.

incomplete metamorphosis—the type of life cycle followed by dragonflies, grasshoppers, and other more primitive insects in which eggs hatch into immature nymphs, which develop larger versions of themselves (instars) before becoming adults, typically with wings and reproductive organs.

instar—a stage in a larva's development. Before becoming an adult, insects often develop into several successive instars, each larger than the previous one. The final instar usually has grown reproductive organs and wings. To create a new instar, the insect has to shed its skin (exoskeleton) and grow a new one.

larva—the immature form of an insect that hatches from an egg.

leaf miner—the immature (larval) insects that tunnel themselves between the two surfaces of a leaf. As they mine this area by chewing up and eating the internal tissues there, leaf miners create distinct patterns in the leaves.

maggot—the immature larval form of some flies.

molt—the shedding of an insect's exoskeleton (skin) to make room for new growth.

neuron—a nerve cell transmitting electrical signals.

nuptial gift—food or other offering a male insect gives to a female to entice her to mate with him.

nymph—the immature stage of an insect with incomplete metamorphosis.

ovipositor—the tubelike extension from the rear (abdomen) of an insect that it uses to deposit its eggs. Wasp ovipositors are needlelike and can be used to pierce and then inject eggs into another insect or other substrate. The ovipositors of bees and wasps can also inject venom, which the insect uses to sting another organism.

pheromone—a powerful chemical signal an animal releases into the air. Insects use pheromones to communicate with or control the physiology or behavior of others of their kind.

pupa—the stage of an insect's life cycle in which its larval form transforms into an adult. This stage often takes place hidden inside a cocoon or chrysalis. The pupal stage is found only in those insects undergoing complete metamorphosis.

pupate—to undergo the transformation from the larva to the adult form of an insect, which typically has wings and reproductive organs.

sapwood—the soft, still-living outer wood found under the bark. Water and nutrients flow in sapwood to the aboveground portions of a tree or bush.

species—the lowest taxonomic grouping of an insect with features that distinguish it from other species. The scientific name for an insect is the genus name (capitalized) followed by the species name (lower case). Both are italicized.

synapse—the junction between nerve cells through which electrical signals pass. Information is transmitted and stored via synapses.

Additional Reading

For more information about the insects covered in this book and others, check out these books and websites.

GENERAL

The Book of Bugs: The best writing about insects, arachnids, and other arthropods from Orion *magazine* (*Orion* magazine, 2022)

Bugs in the System by May Berenbaum (Addison-Wesley, 1995)

Bugs Rule! An Introduction to the World of Insects by Whitney Cranshaw and Richard Redak (Princeton University Press, 2013)

Extraordinary Insects: The Fabulous, Indispensable Creatures Who Run Our World by Anne Sverdrup-Thygeson (Simon and Schuster, 2018)

Fabre's Book of Insects by Jean-Henri Fabre (Dodd, Mead and Company, 1921; reprint by Dover Press, 1998)

The Forgotten Pollinators by Stephen Buchmann and Gary Paul Nabhan (Island Press, 1997)

For Love of Insects by Thomas Eisner (Harvard University Press, 2003)

Innumerable Insects: The Story of the Most Diverse and Myriad Animals on Earth by Michael S. Engel (Sterling, 2018)

Insectopedia by Hugh Raffles (Vintage Books, 2011)

Insects Through the Seasons by Gilbert Waldbauer (Harvard University Press, 1996)

Tracks & Sign of Insects and Other Invertebrates by Charley Eiseman and Noah Charney (Stackpole, 2010)

BEES

Bees: Their Vision, Chemical Senses, and Language by Karl von Frisch (Cornell University Press, 1971)

The Bees in Your Backyard by Joseph S. Wilson and Olivia Messinger Carril (Princeton University Press, 2016)

Buzz: The Nature and Necessity of Bees by Thor Hanson (Basic Books, 2019)

The Mind of a Bee by Lars Chittka (Princeton University Press, 2022)

What a Bee Knows: Exploring the Thoughts, Memories, and Personalities of Bees by Stephen Buchmann (Island Press, 2023)

FLIES

The Secret Life of Flies by Erica McAlister (Natural History Museum, London, 2017)

Super Fly: The Unexpected Lives of the World's Most Successful Insects by Jonathan Balcombe (Penguin, 2021)

WASPS

Endless Forms: The Secret World of Wasps by Seirian Sumner (HarperCollins, 2022)

BUTTERFLIES AND MOTHS

Butterfly Gardening: Creating Summer Magic in Your Garden by the Xerces Society and the Smithsonian Institution (Sierra Club, 1998)

Chrysalis: Maria Sibylla Merian and the Secrets of Metamorphosis by Kim Todd (Houghton Mifflin Harcourt, 2007)

ANTS

The Social Conquest of Earth by Edward O. Wilson (Liveright, 2013)

Tales from the Ant World by Edward O. Wilson (Liveright, 2020)

CRICKETS AND KATYDIDS

Crickets and Katydids, Concerts and Solos by Vincent Dethier (Harvard University Press, 1992)

Cricket Radio by John Himmelman (Harvard University Press, 2011)

WEBSITES

songsofinsects.com

xerces.org

Acknowledgments

It takes a village to write and publish a book. I'm grateful to Joy Cutler and Robin Rodriguez for critiquing the first drafts of my chapters, as well as to Workman Publishing editor Maisie Tivnan for adroitly separating the wheat from the chaff in my final manuscript. I also appreciate the careful copyediting done by Nancy Ringer and production editor Samantha Gil, and the creative book design by Katie Benezra. A special thanks to my agent, Alice Martell, who not only expertly shepherded my book into publication but encouraged and advocated for me every step of the way. I'm also grateful to the many experts who took the time to review one or more of my chapters for scientific accuracy, including Jonathan Balcombe, May Berenbaum, Matthew Bertone, Stephen Buchmann, Grzegorz Buczkowski, Jason Chapman, Rex Cocroft, Stephen Deyrup, Charley Eiseman, Douglas Emlen, Michael Engel, Deborah Gordon, Dhruv Grover, Nina Hernandez, John Himmelman, Sarah Kocher, Ted MacRae, Erica McAlister, Rosalind Murray, Seirian Sumner, Douglas Tallamy, and Herb Wilson. Thanks are also due to friends and family who waited patiently while I took photos of any insect I encountered on our hikes. Last but not least, a shout-out to my husband, who always claimed it was fun being married to a nerd whenever I spouted out the fascinating facts I was learning about insects. And, of course, I'm so appreciative of the insects themselves, who inspired me to share their amazing anatomies, superpowers, intelligence, and beauty with readers.

• A fritillary drinks from a verbena flower.

Photo Credits

Alamy: Karel Bock p. 221 (left); BSIP SA p. 121; Scott Camazine p. 175; Ernest Cooper p. 167; Denis Crawford p. 114; Custom Life Science Images p. 221 (right); George Grall p. 50; Imagebroker.com pp. 162, 219; JSK p. 182; Ivan Kuzmin p. 81; Amelia Martin p. 130; Diana Meister p. 172; Dennak Murphy p. 171; Nature Picture Library pp. 88, 141, 166; Jason Ondreicka p. 85; Brian Overcast p. 49; Panther Media Global p. 145; Jeremy Pembrey p. 153; Pixels of Nature pp. 33, 36; Uros Poteko p. 28; Premaphotos p. 113; Bryan Reynolds p. 233; Emanuel Tanjala p. 224; Nick Upton p. 201; Wirestock, Inc. p. 199; Wibke Woyke p. 155. **Creative Commons by 3.0:** p. 39: *Metamorphosis of a Butterfly* by Maria Sibylla Merian (1647–1717). p. 63: Figures 25–32, type specimens of *Catocala*. 25: lectotype, *C. maestosa*, Hulst, 1884. 26: lectotype, *C. manitoba*, Beutenmüller, 1908. 27: lectotype, *C. miranda*, H. Edwards, 1881. 28: lectotype, *C. nebraskae*, Dodge, 1875. 29: lectotype, *C. nerissa*, H. Edwards, 1880. 30: lectotype, *C. nevadensis*, Beutenmüller, 1907. 31: lectotype, *C. nuptula*, Walker, 1858. 32: lectotype, *C. nurus*, Walker, 1858. By Lawrence F. Gall and David C. Hawks. **iNaturalist.com:** forrestshimazu p. 70; liuye p. 61; Nancy Mullin p. 58. **Nature Picture Library:** John Abbott p. 146; Clay Bolt/MYN p. 149 (bottom); Alex Hyde p. 149 (top). **Johann Patlak:** p. 47. **Songsofinsects.com:** p. 238.

Vector art: **Getty Images:** Giorgi Gogitidze pp. 73, 129, 249; Illustrator de la Monde pp. 77, 217; kadirkaba pp. 1, 11, 18, 21, 29, 51, 109, 183, 187, 191, 203, 211; Magnilion pp. 37, 48, 59, 89, 95, 100, 103, 115, 122, 125, 143, 152, 154, 157, 160, 163, 173, 227, 235; Nadiinko pp. 67, 195, 241; Pavlo Ozarchuk pp. 18–19, 48–49, 100–101, 122–123, 152–155, 160–161 (background); Alexey Yaremenko p. 179.

About the Author

Margie Patlak is a science writer, memoirist, and photographer. Her book *More Than Meets the Eye: Exploring Nature and Loss on the Coast of Maine* received an "Outstanding Book" award from the American Society of Journalists and Authors in 2022. Her photo book *Wild and Wondrous: Nature's Artistry on the Coast of Maine* was published in 2023, and her photographs have been featured in several solo and group exhibits. She has also written articles for many newspapers and magazines, including *Discover*, *The Washington Post*, *The Los Angeles Times*, and *The Philadelphia Inquirer*; and her essays have been published in a number of literary journals. Patlak divides her time between Down East Maine and Philadelphia.